# Knowledge and Wonder

# Knowledge and Wonder
# The Natural World As Man Knows It

Second Edition

**Victor F. Weisskopf**

The MIT Press
Cambridge, Massachusetts, and London, England

*Knowledge and Wonder* was originally published in 1963 as part of the Science Study Series developed by Educational Services Incorporated. A revised edition was issued in 1966.

Second MIT Press printing, 1980
Copyright © 1962, 1966, by Doubleday & Company, Incorporated
New material © 1979 by The Massachusetts Institute of Technology

This book was set in VIP Baskerville by Grafacon, Inc., and printed and bound by The Murray Printing Co. in the United States of America

**Library of Congress Cataloging in Publication Data**

Weisskopf, Victor Frederick, 1908-
  Knowledge and wonder, the natural world as man knows it.

  Includes index.
  1. Science—Popular works. I. Title.
Q162.W4  1979    500       79-19148
ISBN 0-262-23098-4 (hard)
      0-262-73052-9 (paper)

. . . for all knowledge and wonder (which is the seed of knowledge) is an impression of pleasure in itself . . .

*Francis Bacon*

# Contents

# Preface

This book had its beginning in a series of lectures the author gave at the Buckingham School in Cambridge, Massachusetts, before an audience with no special grounding in science. The idea was to sketch our present scientific understanding of natural phenomena and to try to show the universality of that understanding and its human significance.

Now such an undertaking runs into difficulties that are only too well known. Scientific knowledge is hard to communicate to the nonscientist; there is so much to be explained before one can come to the essential point. All too often the layman cannot see the forest, but only the trees. The difficulties, however, should not prevent, or even discourage, scientists from tackling the job in different ways. This book is one way of giving the uninitiated an idea of the greatest cultural achievement of our time.

Today the different natural sciences are no longer independent of each other. Chemistry, physics, geology, astronomy, and biology are all linked together, and all are treated in this book, though some at greater length than others. Physics, being the basis of all the natural sciences, gets the main emphasis—in particular, atomic physics, since everything in nature is made of atoms. What is stressed in the book is the trend toward universality in science, from the elementary atomic particle to the living world, a common point of view whose realization seems nearer because of the enormous progress the last few decades have brought in our understanding of atoms, stars, and the living cell.

In writing a book as small as this one, the author must make a selection and necessarily will have to leave out many important topics. The choice was based upon the author's own views of the importance of various fields and, to no small extent, upon his restricted knowledge. There is one omission that needs some comment. Einstein's theory of relativity is not included and is scarcely mentioned. There is no question in the author's mind that the Einstein theory is one of the greatest achievements of physics and of all science. It has revolutionized our ideas of space and time to such an extent that without Einstein no exact quantitative consideration of space and time is possible. Einstein's ideas, therefore, play a decisive role in the *quantitative* formulation of many scientific problems. This book, however, emphasizes the *qualitative* aspects of the picture of the world seen in science. Relativity theory is not absolutely necessary to this view, except the recognition that energy and mass are equivalent. We therefore have left out the discussion of the theory of relativity with the exception of the transformability of mass and energy.

I received help from many people who read the early versions of the manuscript and suggested changes and additions. I am deeply indebted to my fellow scientists David Hawkins, Mervyn Hine, Philip Morrison, Alex Rich, and Cyril Smith. I derived much stimulation from the book *Physics* edited by the Physical Science Study Committee (D. C. Heath and Co., 1960). I also want to express my particular gratitude to two nonscientists, Kingman Brewster and Ann Morrison, for their help as guinea pigs in the early samplings and for their constant encouragement throughout. Special thanks go to Mr. John H. Durston, of Educational Services Incorporated, for his careful revision and his improvements in the manuscript; to Mr. R. Paul Larkin for his illustrations; and to the Buckingham School, whose invitation to lecture brought this book into being.

*Victor F. Weisskopf*
*Geneva, Switzerland*
*March 1, 1962*

**Note to the Second Edition**

The republication of this work by The MIT Press gave me the opportunity to improve many parts of it. In particular, the chapter on chemistry has been largely rewritten, and much has been added to the chapter on life. Also, the section on subnuclear phenomena has been brought up to date. I am most grateful to Mr. Edwin F. Taylor for his help and criticism in this task.

V.F.W.
July 1978

# Knowledge and Wonder

# 1 Our Place in Space

How large is the world? How large are the objects about us in this world? We have an immediate feeling for the size of the objects with which we deal in our daily life. The smallest length our eyes can perceive is the breadth of a hair; it is about a tenth of a millimeter across.[1] The human body is roughly two meters high; that is a little more than ten thousand times larger than the breadth of a hair. Other objects about us, such as furniture, tools, cars, houses, are all of roughly the same size as our body; if this were not so, we could not handle them easily.

When we look out the window at the landscape, we see objects of greater size and distance, such as mountains and plains. We can get an idea of the distances involved by counting the steps necessary to reach them—that is, by comparing them directly with our body. We find that the objects we can see in the distance—the mountains, hills, and forests—are only a few kilometers away, not more than 100 even for the towering Rocky Mountains.

Here ends our direct perception of distance. It would be too hard to measure a continent, not to say the earth, by counting steps. Hence we must use indirect methods for getting an idea

1. We are going to measure all distances in the metric system as all scientists do, and also as ordinary people do in most countries except England and the United States. The introduction of this useful system in Europe is one of the positive effects of the French Revolution; it is deplorable that this custom did not take hold in the English-speaking nations.

The unit of length is a meter, a little more than three feet, approximately the distance between the tip of the nose to the tip of the fingers of an outstretched arm. According to the scientific definition of the founders of this system, the meter should have been one forty millionth of the circumference of the earth. They did not measure things too accurately at that time and made a very small error. We now stick to their original meter. A centimeter is a hundredth of a meter; it is roughly equal to the diameter of a dime. A millimeter is a thousandth of a meter; it is approximately equal to the thickness of a dime. A kilometer, one thousand meters, is a "lean" mile, 3300 feet.

of the sizes and distances larger than, say, 100 kilometers. One method is the measurement of distance by speed. If I travel from one point to another at a given speed, say 100 kilometers per hour, and know the time it takes, I can figure the distance. Modern means of transporation have made it easy. An airplane takes about 6 minutes to travel 100 kilometers; it takes about 300 minutes to fly from coast to coast. Hence the size of our continent is about 5,000 kilometers. It would take the same airplane almost eight times longer to fly around the world; thus, the circumference of the earth should be about 40,000 kilometers. Indeed it has exactly this length. Since we know that the earth is a sphere, it is easy to figure the diameter from the circumference—13,000 kilometers. This is the size of our own abode, the planet Earth.

## The Distance of the Moon, the Sun, and the Planets

Now we turn our attention to the heavenly bodies. How can we measure their distance and size? The sun, the moon, and the stars seem all to be pinned on some domelike surface enclosing the space in which we live. When we look at the starry sky, it is as if all the heavenly bodies were at same distance from us. (See figure 1.) The actual distance of these bodies is too vast to be directly perceived.

But there are very simple ways of measuring the distance of the closer heavenly bodies. The simplest method is a very new one based upon the recent advances in radar technique. One directs a beam of radar at the object and sends off a short signal. One waits for the return of the reflected radar wave and measures the time elapsed between emission and return of the signal. When the signal is aimed at the moon, the elapsed time interval has been found to be 2.6 seconds. It took the radar signal that long to travel to the moon and return.

Figure 1
Woodcut showing the prevailing idea of the world during the Middle Ages.
The traveler puts his head through the vault of the sky and discovers the
complexities that move the stars.

The radar wave is a kind of light wave, and all light travels at
the same speed (see chapter 3)—300,000 kilometers per second. We therefore conclude that the distance earth-moon-
earth is 2.6 × 300,000 kilometers, which tells us that the moon
is about 400,000 kilometers away. This is again a distance
measurement by speed.

And how large is the moon, now that we know its distance?
From the earth we see the moon in the form of a disk. Its size
is such that it would take 360 disks like it, side by side, to make
a big half circle from the horizon west, up to the zenith, and
down to the horizon east. Since we know the distance to the

moon, we also know the length of this half circle whose radius
is the distance to the moon. The circumference of a circle is
$2\pi$ times the radius, where $\pi$ is the well-known number
3.14 . . . . Hence the half circle has a length $\pi$ times the
radius, or $\pi$ times 400,000 kilometers. The diameter of the
moon must be 1/360 of this length, and this is about 3,500
kilometers. It is a third the size of the earth, and the distance
to the moon is only a little more than thirty times the diameter
of the earth. Our moon is almost a terrestrial object.

Let us now look at other heavenly objects, but first only at
the members of our solar system, the sun and the other
planets. Men have observed the motions of the planets for
many centuries and have wondered what they signified. Since
the days of Copernicus it has been clear that the strange
movements of the planets are the motions of bodies orbiting
in circles (actually ellipses, which are almost circles) around
the sun as we see them from the earth, which also circles the
sun. The earth is one of the planets, the third one when the
orbits are counted from the sun out. Careful observation of
the motions as seen from the earth reveals the relative sizes of
the orbits of different planets. For example, Mercury always
is observed close to the sun and never farther away from it
than 23°; from this fact we conclude that the radius of Mercury's orbit is 0.38, or a bit more than one third, of the radius
of the orbit of the earth. In the same way we find that Venus's
orbit is 0.7 of the earth's or a little more than two thirds. Thus
we can construct a picture of the solar system in the right
proportions, but we do not know its actual size. (See Figure 2.)

How do we find the size of these orbits and then obtain
an idea of the true dimensions of the solar system? Since we
know the positions of all the members of the solar system
relative to each other, we need only to measure the true distance of one of them in order to find the true dimensions of all

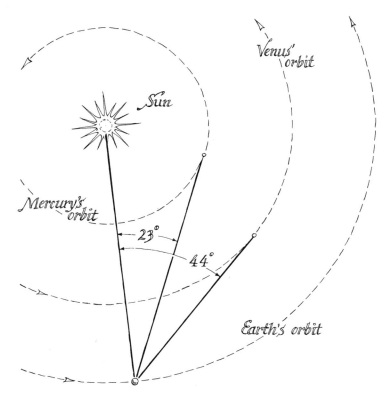

Figure 2
The largest angles at which we see Mercury and Venus away from the sun.
They determine the ratios between the Earth's orbit and the orbits of Mercury and Venus.

orbits. Here again we can use the radar method of measuring distances within the solar system.

Although some promising experiments have been made, we have not yet been able, at this writing, to use the radar beam technique to obtain useful measurements of the distance from the earth to the sun. We can, however, direct our radar beam to one of the nearby planets. This has been done with Venus, and the time between emission and return of the signal was somewhere between five and fifteen minutes, depending upon where in their orbits Earth and Venus happened to be at the times of observation. From the speed of light we can figure out that the distance to Venus is of the order of millions of kilometers.[2] Thus we have found a distance characteristic of the solar system. The size of the solar system is such that light takes minutes to travel from one planet to the other. Once we have determined one single distance such as Venus–Earth, it is no longer difficult to find any other distance in the solar system, since we know the proportions and the relative sizes of the orbits. Right away we can find the distance that is most important for us here on Earth, and that is the distance of 150 million kilometers from the sun to the earth. Light takes a little more than eight minutes to travel from the sun to us.

How big is the sun? It appears to be as large as the moon, but it is, as we can easily figure out, 389 times farther away.

2. In this book, we frequently determine distances, sizes, masses, energies, or other quantities; we will not be interested in the very exact values. For an understanding of nature it is often more important to know the approximate value of the quantity; we want to get a rough idea how big it is. To get a feeling for the size of the earth, it is more revealing to know that the earth's diameter is "about" 10,000 km. than to know the exact value of the diameter (12,756.326 km. on the equator, and 12,713.554 km. between the poles). We therefore use the expressions "order of" or "order of magnitude of" when we want to give a general idea of the size of a quantity.

Hence the diameter of the sun must be 389 times larger than the moon, and that multiplication gives 1.4 million kilometers. The sun is more than a hundred times larger than the earth. (See figure 3.)

## The Distance of the Stars

We have now measured the size of our solar system, which, better than the planet Earth alone, deserves to be called our abode. After all, the sun is our main source of light, warmth, and energy. It is the star to which we belong and it is part of our life. The solar system is the world in which we live. Let us look outside.

All we see there are stars. They are called "fixed stars" because they seem to be immovable in contrast to the planets, whose motion around the sun is plainly visible. Actually the stars are "fixed" only because they are so far away that any motion they might have would be too slow to observe within a lifetime. In fact, they do move. Exact photographs of the sky show slight changes in the positions of stars over periods of many years. We can infer from ancient scriptures that some of the stellar constellations looked quite different many thousand years ago.

But how far away are they? Let us first make the assumption—in reality it is not true—that the stars we see in the sky are all about as luminous and as large as the sun. The stars in the sky do not appear to be equally bright; there are bright ones and weak ones. If our assumption is correct, this can only be due to the fact that some are farther away and some are nearer. We then can easily calculate the distance of a star.

Let us look at Sirius, for example, and take into account the well-known fact that a light source appears weaker when it is farther away. If one of two equally bright light sources is $n$

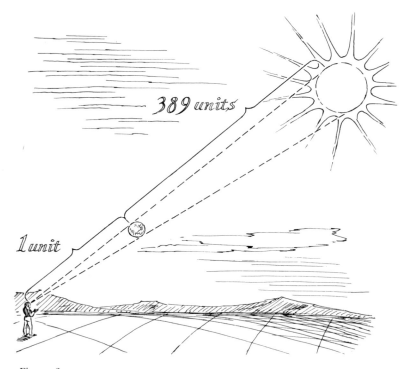

Figure 3
Relationship of moon and sun to Earth observer.

times farther away than the other, the nearer one appears $n^2$ times brighter than the farther one. Let us apply this law to Sirius and the sun. The sun appears much brighter than Sirius. If we compare the light intensities, we find that the sun is 10 billion times brighter than the bright star Sirius. Therefore, if Sirius had the same luminosity as the sun, it would follow from our law that Sirius is a hundred thousand times (the square root of ten billion) farther away than the sun. Actually, Sirius is about 25 times more luminous than the sun. Hence we must place it five times farther away, since this increased distance reduces its apparent brightness by $5^2 = 25$. Thus its distance from the earth is five hundred thousand times that of the sun. Other stars, such as the seven stars of the Big Dipper, appear about nine times weaker than Sirius. So they must be about three times farther away, if we assume that they have the same luminosity as Sirius. It would be easy to find the distances of all stars, and thus the size of our visible universe, if it were correct that most of the stars are about equally luminous.

Can we measure directly the distance of the stars? Yes, we can. The simplest method to measure the distance of an object beyond reach is to look at it from two different points and note the change of direction in which it appears. A distant tree will be seen in a slightly different direction if I walk a few steps in a direction perpendicular to the line between me and the tree. The farther the tree is away from me, the smaller is this change. The distance of the tree can be calculated from that change of direction.[3] Of course the stars are so far away that we cannot observe the slightest change of direction when looking at a star from different locations on Earth. But we can

3. If I express the change of direction in degrees, say $a$ degrees, and if I walk $n$ meters, the distance of the tree is $57 \times n/a$. The smaller $a$, the larger the distance.

make use of the fact that the earth circles around the sun and that we, therefore, are changing our point of observation all the time. (See figure 4.) In the winter, in fact, we are looking at the stars from a spot 300 million kilometers away from the spot where we look at the stars in the summer. When we move from one point of the circle to the opposite point, the stars, in particular the nearer ones, should appear slightly shifted. When the earth has moved, say, from right to left during the half year, the star should appear to have moved from left to right relative to the background. In fact, the star appears to have moved across the sky a distance that equals the diameter of the earth's orbit (300 million kilometers) as the orbit would

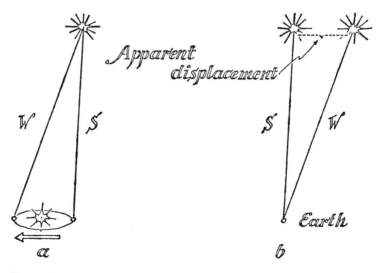

Figure 4
Apparent displacement. In winter the star is seen in the direction *W* and in summer in the direction *S,* as shown in (a). Seen from the earth, the star therefore appears displaced by a distance equal to the diameter of the earth's orbit, as seen in (b).

appear to a viewer as far away as the star. If our previous hypothesis is right, Sirius, being then half a million times farther away than the sun, should be found to perform small periodic displacements not larger than a dime would appear to be when viewed at a distance of 2 kilometers (half a million times the radius of the dime). These displacements actually have been found! For about 120 years the astronomers had instruments to measure such small changes in position. Using this method, they were able to measure the distances of hundreds of stars that were not farther away than 100 light-years. Among them are most of the brighter stars in the starry sky.

We now know the distance of the brightest, and therefore the nearest, stars. We can judge the amount of empty space between our solar system and the nearest sunlike objects: It is about a million times the distance earth–sun, or about $10^{14}$ kilometers.[4] The light takes ten years to travel this distance, and this is why we measure distances of that order in light-years. Sirius is ten light-years away. Compare this with the few minutes that light takes to travel within the solar system, or with the fact that light needs only one tenth of a second to travel around the earth, and you get an idea of the distance to our nearest sister suns.

4. Instead of writing big numbers with many zeros, we shall use in this book the common scientific notation with the so-called "powers of ten." The number $10^{14}$ means ten to the fourteenth power; that is, 10 times 10 times 10 and so on, fourteen times—thus a number with a one as the first digit and fourteen zeros following it. A million, for example, is written as $10^6$.

When we say that Sirius is $10^{14}$ kilometers away, we do not mean that its distance is exactly this number of kilometers. We only have given the "order of magnitude." It might be 2/3 of $10^{14}$ or 1½ times more. For special scientific purposes it is necessary to know this distance much better—it is actually known with very good accuracy—but for our survey it is not necessary to know this figure exactly. It makes no difference for our general insight into the dimensions of the universe whether Sirius is ½ × $10^{14}$ kilometers away or 2 × $10^{14}$ kilometers. We are interested in the order of magnitude of the expanses of space.

There are not many stars whose distances can be measured with the displacement method—only the ones that are relatively near to us, not farther than a hundred light-years. There are about 300 stars within that distance. For most of the other stars the displacement is much too small to be seen. Fortunately, there are a number of other more indirect ways to find out about the distance of stars. We shall not get into the details of these methods. Altogether our measurements have shown that we make no great error by assuming the actual luminosity of all stars to be nearly the same; that is, stars that appear equally bright have the same distance from the earth. In individual cases, such an assumption may lead to big mistakes; we would place very bright stars too near and very weak ones too far away. But on the average it gives a nearly correct picture of the distribution of stars. Hence for a first orientation as to how the stars are distributed, it is sufficient to assume all stars to be equally luminous, about ten times brighter than our sun. The sun turns out to be a relatively weak star, only a tenth as luminous as the average stars. This assumption of equal radiance of all stars will help us to get a general idea of the vast distances of the universe.

**The Distribution of the Stars**
Let us look at the sky and form a picture of the stars in space. We see bright ones and faint ones, many more faint stars than bright stars. At first sight it seems that there is no regularity in the distribution of the stars in the sky. But when we look more systematically, using a pair of field glasses, we notice quite clearly that the fainter stars are by no means evenly distributed over the sky. There are many more faint stars in or near the Milky Way, compared with the regions away from it. When we look with good field glasses in a direction far from the Milky Way, we find a few brighter stars but almost no very

faint ones. In the Milky Way, however, the background is glittering with millions of stars.

What does this mean? It tells us that the stars are not distributed uniformly in space but are concentrated in a flat disk. Our solar system is somewhere in that disk. When we look into the body of the disk, we see many stars and many far away ones that appear faint; when we look perpendicular to the plane of the disk, we see only a few stars and, because of their proximity, they are relatively bright.

How vast is this disk that contains all the stars we can see in the sky? We again can use our hypothesis and measure the apparent brightness of the faintest stars we see when we look in the direction of the disk (the Milky Way) and when we look out of the plane of the disk. For this we need powerful telescopes that can distinguish every single star in the Milky Way. We then can apply our simple method of determining the distances. Here is the result: The faintest stars we see when looking into the disk are about 100 times fainter than those we see when looking out of the disk. Hence the radius of the disk must be about ten times larger than its thickness.[5] The faintest stars in the Milky Way are about a hundred million times weaker than Sirius; so they must be 10,000 times farther away than Sirius, or about a hundred thousand light-years. (See figure 5.)

The spreading of the stars, faint and bright, over the sky has taught us that the stars form a circular disk with a diameter of $10^5$ light-years and a thickness of $10^4$ light-years. Our solar system does not lie exactly in the center; it is about one third of the radius away from the center. The center lies in the direction of the constellation of Sagittarius. On clear

5. A light source, you will recall, seems 100 times weaker when it is 10 times farther away. In general, if it seems $x$ times weaker, it is $\sqrt{x}$ times farther away.

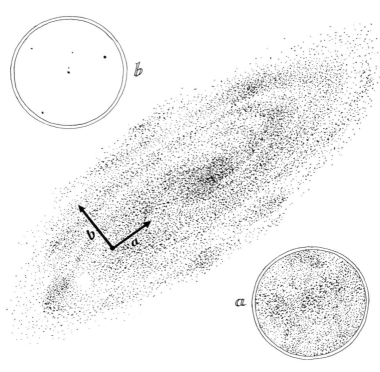

Figure 5
Sketch of the galaxy with an indication of the position of the sun and two directions of view. (a) Stars seen in a telescope looking into the galaxy. (b) Stars seen looking out of the galaxy.

nights we observe that in that region the Milky Way evidently appears to be brightest and most dense.

This flat colony of stars is called a galactic system; it is our own galaxy. The average distance between the stars in this system is of the order of about 10 light-years. This is the distance between the sun and the nearest stars around us and seems to be the normal distance between neighbors in our galaxy. From this we can figure out roughly how many stars there are in our galaxy. We get something of the order of fifty billion.

Nowadays we know considerably more about the structure of our galaxy. It contains not only stars but also gases and dust. This interstellar matter causes difficulties when we apply our simple method of measuring distances. When seen through regions of dust and gas, the stars appear fainter, and we might conclude erroneously that they are farther away than they really are. But the astronomers have worked out many methods to overcome these difficulties. Modern radio astronomy, for example, provides a good method to determine where the interstellar gas is located; this gas emits certain definite radio waves that are characteristic of the atom of hydrogen, the principal element of interstellar gas. By this and other methods (figure 6) we have found that the stars are concentrated in great streamers that form large spiral arms wound around the center in the plane of the disk.

With our increasing knowledge of the structure of the Milky Way, new puzzles occur. The central region of the system seems to conceal many such puzzles. It is surrounded by dust and gas clouds that render investigations difficult. The stars are much closer together there than elsewhere; strange currents of hydrogen gas seem to leak out from the center.

The colony of stars forming our galactic system is the next larger unit in the cosmic environment in which we live. We

Figure 6
Jodrell Bank, a modern radio-astronomical instrument.

first considered the earth and then the solar system as our abode. Now we recognize that the sun, with its planets, is but a small part of a large assembly—many billions of stars—which is our galaxy. What is outside this system?

**Other Galaxies**
Let us again look at the starry sky with our field glasses. We see the billions of stars contained in our galaxy. Once in a while, however, we see something that is not a star; it is a nebula, an extended area of light. A famous and beautiful example is the nebula in the constellation of Orion. This and many other nebulae have been recognized to be big luminous gas clouds. But there are other nebulae—the most striking is

the Andromeda nebula—which in small instruments appear as disk-shaped luminous areas. When these objects are examined with very strong telescopes, one finds that they consist of a very large number of extremely faint stars arranged in the same disk-shaped spiral array as our own galaxy. That was a tremendous discovery! Our own galactic system is not the only one. There exist other similar star systems. The number of these galactic objects is quite large. The more powerful the telescope, the more of these galaxies can be seen. They seem to extend deep into space. How far away are they and how are they distributed?

Again we can get an idea of the distance from the apparent brightness of these objects. Let us look at the Andromeda nebula. Its total brightness is roughly equal to that of an average star about ten light-years away. Very powerful telescopes show that the number of stars in this unit is of a magnitude similar to our own galaxy, about 50 billion. Therefore we must conclude that the nebula actually is 50 billion times brighter than one of our neighbor stars. It appears equally bright; hence its distance must be $\sqrt{50 \text{ billion}}$ greater than the distance of the nearby stars, or 10 light-years times $\sqrt{50 \text{ billion}}$, which gives about 2 million light-years. This number has been checked with other more accurate methods and is roughly correct. The distance between our galaxy and the next one is about twenty times the diameter of our galaxy. The light we see coming from the Andromeda nebula left its source at a time when man had just evolved from his apelike forefathers.

The same method can be applied to the other galaxies we see in the sky. We compare their apparent brightness with the brightness of the Andromeda galaxy and find it, say, $X$ times weaker; we then conclude that the particular galaxy is about $\sqrt{X}$ times farther away. Such a conclusion is based upon the assumption that all galaxies are very roughly equal in size and in actual brightness. There are many detailed observations confirming this assumption.

In this way we can get an idea of the distribution of the many spiral nebulae that are visible with powerful telescopes. There are many millions of them known today. We find them in all directions of the universe. They do not seem to be evenly distributed; they form groups and clusters of galaxies in which they are close to each other and move together in the same direction. On the average, however, the distance between neighboring galaxies is of the order of a few million light-years. The distance between our galaxy and the Andromeda nebula is about equal to the average distance between galaxies. The farther we look, the more galaxies we find. Will this search ever reach an end?

## The Expanding Universe
This question can be answered positively because of a most interesting and unexpected phenomenon that was detected a few decades ago. All these galaxies are moving away from us and away from each other, and the farther away, the faster they move. How do we know?

We must take a closer look at the light that comes from these galaxies. It is the combined light of all the fifty billion stars that make up a galaxy. A light ray is an electromagnetic wave, as we shall see in chapter 3, and the frequency of the wave—that is, the number of complete ups and downs the wave goes through each second—determines the color. We can spread light out into a spectrum by passing it through a prism, and we find the colors arranged according to their frequencies, the lower frequencies on one side, the higher on the other. Now we know from studying starlight that, while all the colors seem to be present, certain frequencies are missing.

When we look at the spectra of most stars, we do find certain frequencies missing. These absent frequencies are the frequencies of the kind of light that is absorbed by the cool gases on the surface of the stars. We find dark lines in the spectrum just at the places where light of these frequencies would be found had there not been any absorption. For example, most of the star spectra have two dark lines in the violet part of the spectrum, indicating the absorption of calcium gas. We are not astonished to find the same two dark lines in the spectra of distant galaxies, since their light is just the sum of the light of all their stars. But it is, or was, startling that these same two dark lines should be found not at the expected frequency but shifted toward lower frequencies. In very distant galaxies this shift is so large that the spots are seen at the red end of the spectrum instead of the violet one.

Such frequency shifts are well known and can be directly interpreted as a consequence of the motion of the object relative to the observer. When a light source moves away from the observer, the frequency of the emitted light appears to become lower, just as the pitch of the horn of an automobile sounds lower if the car moves away from us. The shift is proportional to the speed and, hence, can be used to determine this speed of the object moving away from us.

The observed frequency shift of the light of distant galaxies must be interpreted as a proof of the fact that these galaxies are moving away from us. How fast? It was found that the speed depends on the distance. For the nearby galaxies, such as the Andromeda nebula, the motion is all but unobservable, but galaxies at a distance of about 100 million light-years rush away with the speed of roughly 1,500 kilometers per second. The speed in kilometers per second has been found to be always about fifteen times the distance in million light-years. This relationship between speed and distance was first discovered by the American astronomer Edwin P. Hubble in 1929. At present our largest telescopes can detect galaxies as far away as six billion light-years. They move away from us with a speed of 90,000 kilometers per second, which is about a third the speed of light.

Not only the distance between us and any galaxy increases steadily but also the distance between any two galaxies. The whole universe expands, each galaxy gets farther away from its neighbors, and the density of galaxies decreases steadily.

The expanding motion of the universe of galaxies gives us an indication of the limit of what we can ever see. We might be able to build larger telescopes and try to see galaxies that are still farther away, but they will recede from us with speeds that are nearer and nearer the speed of light. When an object moves away from us with a speed near the speed of light, its radiation will appear to be weakened; in fact, the nearer its speed gets to the speed of light the more the object will fade out, the less visible it will be.

The reason for this is easily understood if you compare the light emitted from a source with bullets shot by a gun in all directions. Obviously the number of hits will seem few and weak from a gun receding from us with a speed almost equal to the speed of the projectiles.[6]

Hence even if there are many more galaxies farther than about 10 billion light-years away (the distance at which the

6. This example might lead one to the erroneous conclusion that the light emitted backward from a source moving away travels at a speed slower than it would if emitted by a stationary source. Obviously bullets fired from a receding gun would come to us at a slower speed than bullets from a fixed gun.

Light always travels at the same speed (300,000 kilometers per second) whether emitted by a fixed or by a moving source. The behavior of light is governed by the laws of relativity, a theory not discussed in this book. The conclusions drawn from our example are still valid however: The light emitted backward by a moving source is weaker not in its speed, but in its intensity. The backward intensity vanishes if the source itself attains the speed of light.

Hubble relation would give a receding speed equal to that of light), even if there are an infinite number of them, we cannot see them; they are moving away from us so fast that their light cannot reach us any more.

A universe of expanding distances between galaxies confronts us with an interesting situation: There might well be an infinite number of galaxies spread over infinite distances. But we can see only the ones that recede from us with velocities reasonably below the speed of light. Hence there is only a finite number of them from which light can reach us. Although this universe might actually be infinite, it is finite as far as we are concerned. We can explore only the part of it that can send us its light signals.

It is remarkable that the astronomical instruments at our disposition today, such as the Mount Palomar telescope (figure 7), already can penetrate into distances at which the speed of recession is one third the speed of light. This is not very much less than the farthest distance we will ever be able to see. If we can penetrate only about three times farther into space, we essentially will have spanned the visible universe. Therefore we are witnessing today a great moment in the development of mankind, comparable to the achievement of Magellan's first voyage around the globe in 1520. At that time the planet Earth was encompassed and the limits of travel on earth were clearly recognized. Today we begin to encompass the limits of penetration into space. We may be beginning to observe the last objects that can be observed.

**Ladder of Distances**
Let us now summarize the sizes of things as we have developed them. We build up a ladder of distances step by step, starting with the smallest distance we can perceive with our naked eyes and ascending to the stars.

The smallest distance we can distinguish is about one tenth

Figure 7
Mount Palomar Observatory.

of a millimeter. It is the width of a hair. The next step on our ladder is a distance characteristic of our own size—the distance from our eye to the tip of our arm; ten thousand times larger than the first one, it is about one meter. The distance to the mountains plainly visible at the horizon is again ten thousand times more: 10 kilometers. The diameter of the earth is the next step; it is about one thousand times more, namely 12,000 kilometers. The distance earth-sun is again about ten thousand times larger; it is 150 million kilometers. The next step of our distance ladder is the distance of the nearest stars; this time the jump is a factor of one million, and we get $10^{14}$ kilometers, or 10 light-years. The following step is the size of our galaxy, which is again ten thousand times the previous distance, namely $10^5$ light-years. The next step is

Distance Ladder

| Distance | | Ratio between steps |
|---|---|---|
| Smallest visible distance | 0.1 mm. | |
| | | 10,000 |
| Human dimensions | 1 m. | |
| | | 10,000 |
| Objects in landscape | 10 km. | |
| | | 1,000 |
| Diameter of Earth | $1.2 \times 10^4$ km. | |
| | | 10,000 |
| Earth–sun | $1.5 \times 10^8$ km. | |
| | | 1,000,000 |
| Sun–Sirius | $10^{14}$ km. | |
| | | 10,000 |
| Size of galaxy | $10^{18}$ km. | |
| | | 10 |
| Distance of near galaxy | $10^{19}$ km. | |
| | | 10,000 |
| Size of universe | $10^{23}$ km. | |

larger only by a factor between ten and one hundred; it leads us to the distance between nearest galaxies, which is several million light-years. The final step, another factor of ten thousand brings us to the distance of the farthest objects that ever can be seen, to what we may call the radius of the accessible universe; according to today's best knowledge, it is somewhere between 10 and 30 billion light-years.

This is the end of our ladder of distances. Each step has led to a greater distance. In most cases the increase was by a factor of ten thousand. Such a step is easily within our ability to visualize by remembering that the length of our arm is about ten thousand times the width of a hair, or that a distance of ten miles is related by the same factor to the length of our own body. Even the factor one million, which occurs between the distance earth–sun and the distance to the next star, can be comprehended: The distance to Sirius is to the distance to the sun as 100 yards is to a hair's breadth. Our intuition begins to falter, however, when we try to grasp the extension of the whole ladder. The tremendous expanses of the visible universe are too great for any immediate comprehension in terms of terrestrial sizes. All the greater is the achievement of the human mind, wherein were created the ideas that have led to the recognition of the vast dimensions of the universe. Blaise Pascal, the great French philosopher, said, "It is not the vastness of the field of stars which deserves our admiration, it is man who has measured it."

# 2 Our Place in Time

## The Age of the Landscape

How old is the world? We have an immediate perception of time intervals pertaining to our life. The shortest time that we can sense directly is about a tenth of a second. It is the duration of a snap of a finger. The natural units of time we deal with in our daily life are the day and the year, and we know what it means when we speak of a man's life span. Recorded history traces back the flow of time for about 5,000 years. This takes us back to the Sumerian civilization, which is the most ancient period known to us from written records. Hence 5,000 years is the longest period for which we have direct human experience. If we want to learn the chronology of events that antecede human history, we must use indirect methods.

The large forms of nature around us—mountains, hills, rivers, oceans, and plains—have not changed much in the course of written history. Have they been here for eternity? Evidently not. Wind and weather are wearing them down.

Let us consider a mountain such as the Matterhorn, which is in the Alps between Italy and Switzerland. It rises above its immediate surroundings about 2,000 meters and, at its base, is about 2,000 meters broad. Hence it contains roughly $2 \times 10^9$ cubic meters of rock. It slopes have an area of roughly $10^7$ square meters. The weather—rain, ice, and storm—breaks off little pieces of rock here and there, mainly by freezing water in cracks, and the mighty structure is slowly destroyed. How long would it take to level it down? Let us perform a simple calculation. It is reasonable to assume that on the average a piece of rock a few inches in size is broken off per year from every square meter. This gives us about $10^3$ cubic meters per year coming down from the Matterhorn. After a million years half of the mountain would be gone. The life of a mountain such as the Matterhorn must be of the order of millions of years.

We come to similar conclusions when we study the amount of material the rivers transport to the seas. We can measure the amount of fine-grained rock, sand, and soil swept from the land by rain and brought to the sea by the rivers in one year. If this material were distributed evenly over the land from which the rivers come, it would be a very thin layer, only 1/300 of a centimeter thick. In a million years, however, it would be a layer of thirty meters. Since the material does not come evenly from all points, but only from spots where there is a slope, we see that rain and weather in one million years can remove hills many hundred meters high and change our landscape appreciably. So the age of the landscape we see around us can be counted in millions of years.

Erosion by rain and wind is the destructive leveling force that has shaped the surface of the earth. If there were no other force acting, the world would all be flat, since mountains and hills would have been erased in a few million years. But there are constructive forces at work that slowly but constantly change the surface of the earth. The inside of the earth is under high pressure, since it sustains the whole weight of the outer layers. This pressure sometimes is released at one point, or increases at another. The pressure changes cause movements of the surface up and down; high plateaus and deep depressions are formed. Sometimes the movements are sideways, and the surface curls up to form ridges and valleys, just as a piece of cloth does when it is pushed inward from two opposite sides. (See figure 8.) There is a constant interplay of mountain formation and subsequent destruction by erosion. We live in a period separated by only a few million years from a time of very violent mountain formations; this is why the surface of the earth now exhibits so many

Figure 8
Surface of Earth like piece of clo.n.

different mountain ranges. Some fifty million years from now the earth might be much flatter and less interesting if no new mountain-forming events should occur in the meantime.

## Radioactivity: The Clock of the Universe

How long did this constant interplay between mountain formation and atmospheric mountain destruction go on? How can we measure the time intervals in which the great geological events occurred? We must use a natural clock that turns slowly enough to measure long times in a way we can read. Fortunately, nature has provided a very slow, regular process that can be used for time measurements. It is radioactivity, the strange phenomenon discovered in 1896 by the Frenchman Henri Becquerel. But how can radioactivity be used as a clock?

When it was discovered, radioactivity puzzled everyone because it disproved the old belief that chemical elements are unchangeable.[7] The phenomenon of radioactivity shows that

7. Chemical elements are the materials of which all forms of matter are composed, such as iron, gold, oxygen, sulphur, and carbon. We will learn much more about elements in chapter 4. The phenomenon of radioactivity is discussed in more detail in chapter 7.

some elements are not. A radioactive substance changes over into another substance. The atoms of such a substance undergo a decay with the emission of rays and become the atoms of another element.

Let us consider an example, radioactive rubidium. Rubidium is a relatively rare metallic element, somewhat similar to potassium or sodium. There are two kinds of rubidium found in nature (two "isotopes"). They differ in weight; one has the atomic weight 85, the other 87, and the heavier one is a radioactive element.[8] A piece of pure $Rb^{87}$ emits a characteristic radiation whose nature is not important to us here. (It is quite important to the medical profession, since it can be used as a treatment of cancer.) The main point is the fact that one atom of $Rb^{87}$ changes into a different kind of atom, into an atom of strontium. This transformation occurs slowly and steadily at a fixed rate in time that cannot be accelerated or slowed down by any outside influence. Every year a certain fraction of $Rb^{87}$ is transformed into strontium. This fraction is extremely small, only $1.6 \times 10^{-11}$ per year.[9] Thus every year a hundredth of a billionth of $Rb^{87}$ changes into strontium. Most of the radioactive substances found in nature transform as slowly as this. The "decay constant" of uranium is $2 \times 10^{-10}$ per year, which means that only two parts in ten billion ($10^{10}$) change every year. Potassium 40, also a radioactive element, has a decay constant of $0.7 \times 10^{-9}$ per year.

Our knowledge of radioactivity has increased greatly since

8. The atomic weight is the weight of an atom compared to the weight of the lightest atom, the hydrogen atom. Rubidium 85 has an atom that weighs 85 times as much as the hydrogen atom.

9. Here we use the method of negative powers of ten in order to express very small numbers. The exponent $10^{-1}$ means $1/10$; $10^{-2}$ means $1/10 \times 1/10 = 1/100$; $10^{-11}$ means $1/10 \times 1/10 \times$ . . . eleven times. One can also say that $10^{-11} = 0.00$ . . . 1 with eleven zeros, including the first zero before the decimal point.

the invention of high energy accelerators (atom smashers). In these machines small particles are hurled with high energies against the atoms of various substances, and they produce changes in those atoms. For example, normally nonradioactive elements are changed under bombardment in these machines into new elements, which in most cases are not found in nature and very often are radioactive. Thus one can produce new "artificial" man-made radioactive materials of great value in physical and medical research. Most of them transform very much faster than the natural radioactive elements. For example, one can produce a radioactive sodium (weight 24) that transforms into magnesium at the rate of 6 percent per hour.

## The Age of Matter

Now we come to our first fundamental conclusion about cosmic time scales: The earth could not have existed forever. There are things on Earth that could not have been here forever. If the earth had existed for an infinite time, we could not find naturally radioactive substances on its surface, such as $Rb^{87}$, uranium, or potassium. In fact, if the age of the earth were much greater than $10^{10}$ years, all the naturally occurring radioactive substances we have mentioned would have transformed almost completely into their daughter products and would not be found. We know well that the process that made those elements no longer continues.

How old, then, is the earth? When we look at the decay constants of the radioactive substances found in nature, we observe that they are always less than one billionth per year. The artificial radioactive materials, however—the elements that we produce ourselves—have all kinds of decay constants. They range from slow decays, such as a fraction of a millionth per year, to really fast decays, such as the decay of half the material in a few tenths of a second. Examples of all decay speeds between these limits have been found, but only those decaying more slowly than a billionth in a year are ever seen occurring in nature. The explanation is quite simple. The faster ones cannot be found since they already have disappeared in the time the earth has existed.[10]

From this we conclude that the matter of the earth must have existed in its present state for several billion years, but not much longer. The naturally radioactive element with the fastest decay is uranium 235[11] (one billionth per year) and already it is all but gone; it is found only as a very small percentage (0.71 percent) of ordinary uranium. The latter is mostly uranium 238, which decays more slowly, about 0.15 billionth per year. The "age" of the matter that composes our Earth must therefore be somewhat longer than one billion years, perhaps five or ten times longer, but not much more.

It was an impressive moment in the history of our scientific recognition of the world when proof was found, here on Earth, that the earth had not existed forever. The radioactive materials are only a tiny part of the earth. They are extremely rare. Still their very existence bears witness to a beginning of some sort.

What happened at that beginning? Evidently the earth could not have been in a state resembling the present situa-

10. There are interesting exceptions, which show how easily a false clue can fool one. In fact there exist a few fast-decaying radioactive elements in nature. However, they are all what we call "daughter substances" of slowly decaying radioactive elements. Here is what this means. It happens sometimes that the product of a radioactive decay is again radioactive and also decays into a third element. Such a product is called a "daughter substance." If the first decay is very slow, and the second is fast, we observer the fast decay in nature following the slow one every time.

11. Uranium 235 or 238 means that kind of uranium whose atoms weigh 235 or 238 times the weight of a hydrogen atom. One refers to them also as $U^{235}$ or $U^{238}$.

tion. At that time the material of which the earth is made must have been subject to conditions under which radioactive elements could have been created. These are the conditions that we produce in our big nuclear accelerators. Particles and atoms must have been moving with tremendous energies and at high densities, colliding with each other at great speed. The temperatures necessary to produce these conditions range in the region of 100 million degrees. We have good reasons to believe that such conditions occur in the center of stars, not in ordinary circumstances but when stars become unstable and explode. Exploding stars are called supernovae; they appear suddenly as new stars in the sky and fade away in a few months. They are nothing very unusual. There should be about twenty or thirty per year among the fifty billion stars of a galaxy.[12]

Hence we come to the conclusion that the material of which the earth is made must have been subjected to tremendous heat and acceleration, probably from exploding stars, at a time that lies roughly five to ten billion years in the past. We can consider these events as the creation of the elements of which our environment is made.[13] At that time many radioactive and nonradioactive elements were created, including all those we can make with our nuclear machines and certainly a few more. But the radioactive substances with shorter lifetimes have long since decayed and are transformed into stable elements. The few long-lived, naturally radioactive substances are the last witnesses of the eventful times at which the elements were formed—the elements that now make up matter on Earth. They are the last embers still remaining from

12. The chances that an ordinary star such as the sun will explode are not very high. It happens only once in a few billion years. The last explosion in our neighborhood (within 1,000 light-years) occurred in 1750.

13. Chapter 9 contains a discussion of element formation in stars.

the great cosmic fire that, ten billion years ago, created the materials we see around us on Earth.

## The Dating of Events in the Earth's History

A closer analysis of the decay process of natural radioactivity allows us to determine other events more recent than the birth of our elements. A mineral containing a naturally radioactive product also contains the decay product of radioactivity in most cases. For example, in a rock containing rubidium, one also finds strontium, the element into which the radioactive rubidium transforms. By comparing the relative amounts of the two elements, one can calculate how long the rubidium was kept in this rock, or, in other words, one can calculate the time that has elapsed since that piece of rock solidified. The calculation is quite simple. Every year $1.6 \times 10^{-11}$ parts of the rubidium turn into strontium; one can infer, therefore, how many years it took to get the observed amount of strontium.

There is only one hitch in it. Not all the strontium in the rock need have been rubidium before. The rock might have picked up original strontium too. There is a very elegant method to avoid this difficulty. The strontium from rubidium is a very special one, namely $Sr^{87}$. Normal strontium consists only of 12 per cent $Sr^{87}$ and the bulk is another strontium isotope, $Sr^{88}$.[14] Hence all one has to do is measure also the amount of $Sr^{88}$ in the rock. If none is found, then all $Sr^{87}$ must have come from radioactive rubidum. If there is some, then we know how much ordinary strontium was mixed into the rock, and we can determine how much $Sr^{87}$ was added by radioactive decay.

14. The symbol $Sr^{87}$ or $Sr^{88}$ means a strontium whose atoms weigh 87 or 88 times the weight of a hydrogen atom, respectively. The number 87 or 88 is the atomic weight of strontium.

Similar measurements can be made with rocks containing any other natural radioactivity. The radioactivity of potassium and uranium have been used widely for this purpose.

The naturally radioactive materials do more than bear witness of the origin of the earth; they also can be used as time markers when they slowly tick away with their regular decay.

The measurement of the radioactive-decay products, such as strontium from rubidium, has made it possible to determine the time of events that occurred in the billions of years since our elements were formed. Whenever new mountain ranges appeared, whenever the seas deposited sediment on their floors, the radioactive elements contained in the material began to accumulate their decay products, and we can time the event by measuring the amounts of accumulation. We then obtain a reasonably accurate timetable of geological events. We find, for example, that the Alps and the Himalayan mountain ranges are quite young, only a few million years old. (See figure 9.) The Rocky Mountains in their present form are older, about 60 to 100 million years. The flat ridges of the Appalachian range are as old as 250 to 300 million years, although the actual shapes of the hills on the surface have changed many times since then.

The oldest rocks found so far have an age of 2.6 billion years. Hence the age of the earth must be at least as great. The earth is probably older than this, but what is now the surface of the planet has changed so much that no rock is found today that could show greater age.

The different layers of rocks and sediment include fossils of animals and plants; so the timetable of geological formations leads directly to a timetable of the development of life. (See figure 10.) We find that the earliest records of life start about 600 million years ago with fossils of algae and sponges. Evidently there must have been life earlier in a more primi-

Figure 9
K2 in the Himalayas (28,250 feet) considered the second highest peak in the world.

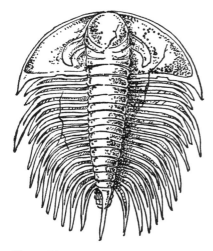

Figure 10
Trilobite (*Olenellus*), Cambrian Period, one of the best-known fossil forms in the study of the earth's history.

tive form that has left no record in rocks. It is estimated that primitive bacteria must have existed as far back as three billion years. Plants and animals with soft tissues that did not leave any traces probably date back to one or two billion years. Fish developed about half a billion years ago, and reptiles began 300 million years ago, when trees and forests began to grow. Birds and mammals developed since roughly 200 million years ago, and man has existed not more than two million years. Thus the radioactive clocks also have helped to date the development of life.

## The Age of the Earth and the Planets

Now and then a piece of matter enters our atmosphere from outer space. Most of these objects—they are called meteorites—evaporate when they enter the atmosphere because of the intense heat created as they speed through the air. Some larger chunks, however, do reach the surface of the earth intact. (See figure 11.) These objects and the paths that led them to us have been studied. It is probable that the meteorites do not come from very far. They are possibly the broken remnants of one or several small planets that disintegrated some time at the beginning of the history of the solar system. The age of these rocks must be close to the age of the solar system itself; they probably date from the time when matter assembled in the form of the sun and the planets. Hence when we can determine the age of these fragments, we probably have measured the age of our solar system, or the time when the planets were formed. Fortunately, the meteorites sometimes contain traces of a radioactive material and its decay products. The amount of the decay product can be used to determine how long the radioactive material decayed inside the piece of matter from outer space. The result has been very consistent: All meteorites seem to be of the same age, an age of 4.5 billion years. We must conclude that this time has passed since matter accumulated into planets in our solar system. Very probably this is also the time our Earth and the other planets have existed as big globes circling the sun.

**The Age of the Stars**
Can we determine the age of other stars? We no longer can use the radioactive clock, since no material reaches us from outside the solar system. Our only means of communication is the light. Still, in spite of the lack of any direct contact, astronomers have tried to get some indirect information about the age of stars. Sometimes it is possible to come to some tentative conclusions by investigating carefully the color and the brightness of stars and by using our new ideas about the processes that produce all the immense energy required to

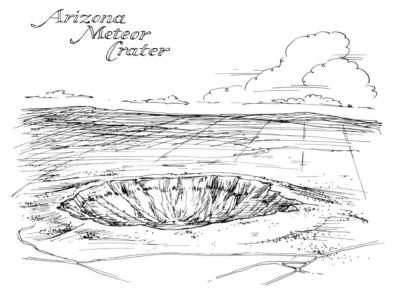

Figure 11
Arizona Meteor Crater. Fell some tens of thousands of years ago. About 1,360 yards in diameter and about 650 feet deep. Probably blasted by a tremendous meteor of iron and nickel, which is probably 400 feet below the floor of the crater and over to one side.

keep a star shining and hot for long periods. (See chapter 7.) From such arguments one finds that the life spans of most stars are measured also in billions of years. Some star groups are possibly as old as twenty or thirty billion years; some very bright stars are only a few million years old.

Just as the year is a suitable unit for the measure of human life, a billion years is the suitable unit for the life of stars. The oldest rocks on Earth are 2.6 billion years old; the solar system must have been formed 4.5 billion years ago; our naturally radioactive substances were formed in some star explosion 5 or 10 billion years ago, and the age of most stars seems also to lie in this interval.

### The Age of the Universe

The expansion of the universe we discussed in the last chapter also gives us a time scale by which the universe develops. Today we see the galaxies steadily streaming away from us, the faster, the farther away they are. We then can ask the question: If it is true that the galaxies move apart from each other, must there not have been a time when they were nearer to one another? In fact, if the expansion of the universe has gone on at the same rate as it goes now, there must have been a time when all galaxies were crowded close together at the same place. Hence the expansion of the galaxies we see could not have gone on forever. When was that time of crowding? We recall that the speed of a galaxy away from us is about fifteen kilometers per second for every million light-years of distance. It is most significant that it takes 20 billion years to go one million light-years at a speed of fifteen kilometers per second. This means that the expansion of the galaxies could not have gone on at the same rate for longer than 20 billion years. Tracing the expansion backward, we must conclude that at that time all galaxies and all matter

Time Ladder

| | |
|---|---|
| Shortest time interval distinguishable by ear | 0.1 second |
| Day | $10^5$ seconds |
| Human life | $10^9$ seconds = 100 years |
| Human civilization | $10^4$ years |
| Development of man | |
| Significant change of Earth surface | $10^6$ years |
| Development of mammals | |
| Age of Rocky Mountains | $10^8$ years |
| Development of life | |
| Age of many rocks | $2 \times 10^9$ years |
| Age of Earth | $4.5 \times 10^9$ years |
| Age of the matter our nearest stars (including sun, earth) are made of | $5-10 \times 10^9$ years |
| Age of universe | $10-20 \times 10^9$ years |

contained in them would have been concentrated roughly at the same place. That must have been the beginning of the expansion of the universe. Again we find a time similar to other cosmic lifetimes.

What is the meaning of all this? Most of the matter contained in the universe today is found in widely separated islands of matter, the galaxies. About 20 billion years ago it could not have been so. All the separate islands must have been connected and pressed together. Very high densities, and therefore high temperatures, must have been the order of the day, and matter must have been under very different circumstances than we find it today. If we were allowed to extend our knowledge of matter that far into the past, and to so uncommon circumstances, we would conclude that things must have been extremely hot, compressed, and turbulent;

the expansion of the universe must have started with very hot matter under high concentration, expanding like an explosion, and pushing matter and light from everywhere into all directions. It is often referred to as the "Big Bang." Some optical reverberations of that bang have been observed today in the form of a "cool" radiation filling all the space. We will come back to it in chapter 9. All we want to say here is that the world as we know it here and now, with its matter, its stars and planets, its galaxies and clusters of galaxies, has existed for about 15 to 20 billion years. Nobody knows exactly in what state the world was before. It may have been a concentrated hot expanse of cosmic fire in which matter was still in some primordial form. All we know for sure is that matter has assumed more or less its present form only for the last 15 to 20 billion years.

If the age of the universe is taken as one day, mankind has existed only for the last ten seconds.

# 3 Two Forces of Nature

In chapters 1 and 2 we set the stage in time and space for what goes on in nature. We now turn our attention to the events on this stage. Here we find an overwhelming multitude of objects, in constant change and motion in the sky and on Earth, with varying properties and qualities ranging all the way from simple gases, liquids, and solids to such intricate aggregates as plants, animals, and men. The behavior of all forms of matter is highly complicated and bewildering. Still we can perceive some kind of order in nature. In spite of constant change and movement, we recognize similarities between different objects; we range them into classes and we have names for them. The materials from which they are made can be classified into definite types, such as rocks, metals, liquids, organic substances, etc. These substances differ widely in their properties, but we observe everywhere the same kinds of metals, the same kinds of rock, the same kinds of organic material, and so forth. A piece of gold is the same wherever on Earth it is found. In the living world, too, we recognize similarities and identities. They are strikingly embodied in what we call the different species; we find bacteria, trees, flowers, and animals that have common properties and are identifiable as being of the same kind.

These are the regularities we want to understand. We want to know why nature has specific forms, why the forms are such and not otherwise, and why the objects behave as we see them behave. To begin with, however, we must first look for simple features in nature that are unspecific and common to all objects. This chapter is devoted to two of these features. One is the phenomenon of gravity and the other is light.

## Gravity on Earth and in the Sky

Gravity is a well-known phenomenon here on Earth. All things around us, large or small, are attracted by the earth—

they fall downward when they are not held up by some support. The attraction of every piece of matter by the earth is the best-known example of a force in nature. Still, tremendous effort and centuries of thinking were needed before mankind recognized that the motion of the moon around the earth and of the planets around the sun is based upon the same force. It long was thought that the laws governing heavenly bodies were different from those that held on Earth. The universality of the laws of nature, their validity for the whole universe, has been recognized only since the days of Isaac Newton.

The moon and the planets do not fall toward the earth nor toward the sun. How, then, could their motion be governed by the force of gravity? There is a big gap between our terrestrial experience of things falling toward the earth and the heavenly appearance of bodies orbiting around a center (moon around earth, planets around sun). The bridging of this gap was a decisive step toward the understanding of the universe. Let us see how it came about.

· Imagine that we are at the top of a very high tower and throw a stone horizontally into space. (See figure 12.) The stone's path will be bent down toward the earth because of gravity, and the stone will hit the ground at a certain distance away from the tower. The harder we throw the stone, the more gradual will be the bending of the path. We can imagine that the stone could be thrown with such vigor that the downward bend of its path would just equal the curvature of the earth's surface, which is, of course, the surface of a sphere. Then the stone would never reach the surface because whenever its path bent down, the surface of the earth would bend by the same amount. We have thrown the stone, as it were, beyond the horizon. If the air did not slow it down, our stone would circle the earth as a satellite. This is, of

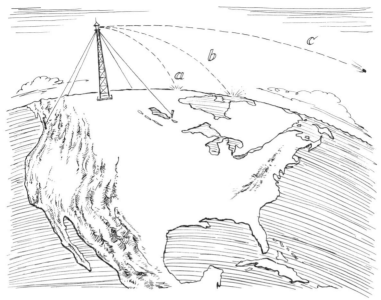

Figure 12
A stone thrown from a tower. Paths *a*, *b*, and *c* correspond to throws of increasing power. Throw *c* will never reach the earth.

course, the principle of launching a rocket satellite. In a typical rocket firing, the first stage raises the satellite above the atmosphere, and then a second rocket explosion pushes it into a horizontal motion. The horizontal speed necessary for the bending to be equal to the earth's curvature is about five miles per second. Thus we see how the falling motion of an object can go over into an orbiting motion around the earth if the object receives a strong horizontal push.

Let us now look at the orbit of a body around a center of attraction in a different way. When a planet circles the sun, the attractive force of gravity keeps the orbit circular, just as a weight tied to the end of a string keeps moving on a circle if you whirl it around while holding the other end of the string. The attractive force counteracts the centrifugal force, which in a circular motion pushes things outward.

The centrifugal force (the drag on the string) is greater the more rounds the object makes per second. It is also greater for larger radii, and it is proportional to the mass of the object. In fact, we can easily calculate the centrifugal force on each planet, since we know its revolution time and its distance from the sun.

The centrifugal force is just balanced by the attractive force of gravity; hence whenever we calculate a centrifugal force in an orbit, we have determined the force of gravity. That is the way Newton measured the force of gravity of the sun upon the planets and of the planets upon their moons. He found that gravity follows a very simple law: The attraction between two bodies is proportional to the product of the masses and inversely proportional to the square of their distance. For example, the distance of Venus from the sun is 0.7 times the distance of the earth from the sun. The square of 0.7 is ½, and from this we conclude that the sun's attraction[15] for Venus has to be twice as strong as the sun's attraction for the earth. Here we have calculated and measured a force far, far beyond direct human experience, a force in our sky.

In order to be sure that the force between the sun and the planets is a universal force that acts between any two masses, one must show that the same kind of attraction exists between two blocks of lead or any other two objects, and that this force also decreases with the square of the distance and is proportional to the product of the masses. Of course, the gravita-

15. We need not be concerned here about the masses of Earth and Venus, which happen to be about equal; the mass terms cancel out in the calculation.

tional force between two lead blocks would be extremely small, since their mass is small compared to that of the heavenly bodies. If the blocks weigh one hundred pounds each, the force between them at a distance of one foot is about as small as the gravity force that the earth exerts upon one ten-thousandth of a gram. Still, it has been measured, and these measurements do bear out the general validity and universality of the law of gravity.

## The Generality of the Law of Gravity

Newton's discovery of the law of gravity explained for us the orbits of the planets around the sun. But it also put an end to the old and cherished dream of many philosophers. The dream was to find a fundamental significance in the actual sizes of the orbits and the durations of the periods of the planets. One might have expected that the radii sizes of the planetary orbits would have simple relations; for example, the size should always be doubled from one planet to the next or exhibit some other simple numerical regularity. The Pythagorean philosophers, for instance, attributed special importance to the numerical ratios between heavenly orbits and considered them as the essence of their system. These relationships were the embodiment of a "harmony of the spheres"; they were supposed to reflect an inherent symmetry of the heavenly world as contrasted to the earthly world, which is full of disorder and without any symmetry. The harmonious interplay of the various celestial motions was supposed to produce a music whose chords were audible to the intellectual ear, a manifestation of the divine order of the universe. Even Johannes Kepler, whose analysis of planetary motions led to the discovery of the law of gravitation, tried hard to explain the observed sizes of the orbits by inventing a universe of regular solids—the sphere, cube, tetrahedron, etc.—one inscribed in the next, and each determining the size of one of the orbits by virtue of some deep, fundamental, all-embracing principle. (See figure 13.)

With Newton all these ideas turned out to be illusions. The fundamental principle underlying planetary motion is the law of gravitational attraction. It determines the orbits of the planets only insofar as it requires them to be circles or ellipses with the sun in the center of the circles or in one of the foci of each ellipse, and establishes a special relationship between the radius (or major axis of the ellipse) and the period of revolution. But the principle does not prescribe any special size or radius. In fact, the actual size of an orbit depends on the conditions at the beginning, when the solar system was formed, and on the subsequent perturbations upon the orbits. For example, if initially the earth had received a different speed, it would have circled on a different orbit. Furthermore, if another star should pass near our solar system, all planetary orbits would be changed, and the relationships between their sizes and periods would be quite different after the encounter.

We can see from this that the orbit sizes, as observed today, are of no great significance. They could just as well be quite different without violating any law of physics. The fundamental law of gravity determines only the general character of the phenomenon. It admits a continuous variety of realizations. The sizes of the present orbits in the solar system were determined by the circumstances and special conditions prevailing during the formation of the solar system, or to the influence of passing stars, but there is nothing fundamental in their present magnitudes. We expect the planets of another star to circle on quite different orbits, even if the star is very similar to our sun in its size and constitution.

Because of its universality, the force of gravity reaches be-

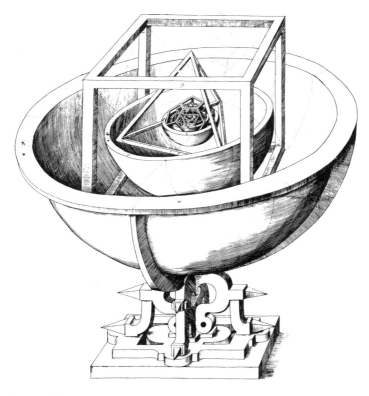

Figure 13
Kepler Device. Kepler's model of the universe, showing how he thought all the planets were positioned with relation to certain geometric shapes. Redrawn from *Mysterium Cosmographicum* (1597, edition of 1620).

yond the solar system and even beyond our galaxy. The stars
within each galaxy attract each other by virtue of gravity, and
each galaxy exerts gravitational forces upon other galaxies;
hence the motions of the stars and also the motions of the
galaxies are regulated by their mutual attraction. We don't
know enough yet about these motions because they are very
hard to observe. We would have to solve a very difficult problem of mathematical analysis if we were to find out what motions an assembly of fifty billion stars would perform under
the influence of mutual gravitational attraction. There are
very good indications, however, that the same principle governs the motions of the stars. The stars seem to circle around
the center of the galaxy in much the same way the planets
circle around the sun.

Are the motions of the galaxies also determined by gravity
forces? Here we come to an unsolved problem of astronomy.
We don't know much about it except for the striking motion
of the galaxies away from each other—the expanding universe. This motion obviously cannot come from gravity; there
must be some other fundamental, but as yet unknown, explanation connected with the Big Bang explosion, which may
have been the beginning of the universe.

## Light

Is there anything more universal than light? The light coming
from the sun to the earth is the basis of our existence. It
furnishes warmth and is the source of almost all energy on
our planet. It makes the plants grow, and we use the plants as
fuel in the form of coal or oil or as food for man and animal.
The only sources of energy that do not come from the light of
the sun are the "dark" forces of radioactivity and uranium
fission. Last but not least, it is in the bright light of the sun that
nature appears to us in all her beauty.

Light is our only messenger from the stars, as Galileo said; it must tell us almost everything we shall ever know about the universe. Except for the meager information we get from cosmic rays and meteorites, and whatever we may learn some day from interstellar travel, we have no communication other than by light with the world outside our Earth.

What is light? The answer to this question is found in one of the last century's most interesting developments in physics. Light signals travel through empty space in straight lines with a fixed speed of $3 \times 10^5$ kilometers per second. In the time it takes to snap your finger (one tenth of a second) light traverses a distance equal to the trip around the earth. As we learned in the previous chapter, it takes only minutes for light to travel between the planets and the sun within our solar system.

When a light signal is sent from one point to another, what is it that goes from the source to the recipient? At first people naturally believed that the light source ejects some light units, impulses or particles, different kinds for different colors. Even the great Newton thought that light consisted of particles (though he cautiously hedged his belief). It was Christian Huygens in Holland who first suggested, in the seventeenth century, that light is a wave motion. Thomas Young and Augustin Fresnel, at the beginning of the nineteenth century, established beyond doubt that a light beam is a wave traveling through space.

What, we now must ask, is a wave? The most common examples are water waves, but they are not the best illustration for the understanding of light waves because they travel on the surface of the water, whereas light waves go through all space. Still, water waves are instructive for the understanding of wave nature.

A wave moves in a carrier. The surface of the water is the

carrier of water waves. The carrier undergoes oscillatory periodic changes; the surface of the water, for example, moves up and down. These changes are so arranged that they move along and make up the characteristic pattern of a traveling wave. We must realize that there is nothing material that travels with the wave. Only the changes of the surface pattern move along as a wave travels. No water is actually transported. Still a wave can transmit effects from one place to another. When I push the surface of water at one side of a pond away from the bank with a plank, the resulting wave transmits this push to the other side of the pond. Water waves can transmit large amounts of power, as we often witness in the effects of waves at the seashore. But the water itself does not move bodily along with the wave. It only moves up and down, back and forth.

Another example, closer to light, is the sound wave. The carrier of sound is the air. The oscillatory changes that the carrier undergoes are air pressure changes. When the sound is produced, say, by a loud speaker, the surface of the speaker moves back and forth, thus producing periodic ups and downs in the air pressure near by. These ups and downs move on in all directions, just as waves on a water surface when you move your hand back and forth in the water. In the water, however, the wave spreads only over the surface; in the air it spreads in all directions of space. This spread of periodic squeezes and releases is a sound wave. When the oscillation reaches the ear, it transmits the pressure to the ear drum, which is put into the same vibrations the source has performed. This pattern of vibration is perceived as sound. The shorter the distance between the ups and downs of air pressure, or the more frequently squeezes and releases alternate when the wave arrives at the ear, the higher the pitch we perceive. The distance between subsequent ups (or downs) is called the wavelength, and the number of ups arriving at the ear (or passing by a given point) per second is called the frequency of the wave. The shorter the wavelength, the higher the frequency.

Although he did not have many facts to go on, Huygens anticipated this idea of light as a wave motion in 1680. The final recognition of light as a wave occurred to an English scientist who started as a medical man, Thomas Young, born in 1773. He worked on problems of light from 1800 on, and he was the first to find the decisive facts that show light as a wave phenomenon.

## Why Light Is a Wave

The vibrations occurring in a light wave cannot be seen directly; only indirect evidence can convince us of the wave nature of light. The best proof today is still the argument Young gave. He based his reasoning upon the phenomenon of interference. Interference occurs when two waves meet each other; it then may happen that the wave crests of one wave coincides with the wave troughs of the other. Whenever and wherever this happens, the wave motion is reduced. Dabble a finger of each hand in water and very closely watch the two waves penetrating each other. You will see that the wave motion is quenched wherever a crest falls upon a trough. If light is a wave phenomenon, light added to light would give darkness under the conditions of such interference.

Indeed, such effects do occur with light. There are many ways in which the interference of light wave can be demonstrated. One well-known effect is the colored bands and rings you see when a thin film of oil is spread over a surface. The colors are often visible at the edges of a patch of oil on a street pavement. Here light from the sky or from a street light is reflected, first at the upper surface of the oil film and then at the lower surface. The beam of light reflected at the lower surface is behind the beam reflected at the upper surface by a distance of twice the thickness of the film. The two reflected beams "interfere" in the following way: If the thickness of the oil film is a quarter of a wavelength, the second is behind the first by half a wavelength. The crests of the wave reflected from one surface fall upon the troughs of the wave reflected from the other one, and we get darkness. This interference causes the white daylight to become colored upon reflection; white is a combination of all colors. Certain colors may have just the wavelength that gives darkness upon reflection. The reflected light then appears in the hue of the remaining colors.[16]

You can observe this interference effect in a simple experiment. Hold a phonograph record at eye level with a lamp in the background, in such a position that the light from the lamp strikes the flat record at a very small angle. You will see a pattern of color on the record edge near your eye. Light beams reflected from different grooves on the record interfere with one another, causing bands of darkness and bands of bright color.

Another example of interference effects is shown in figure 14, which illustrates what happens when a sharp edge casts a shadow on a screen. Light waves are scattered at the edge of the barrier. Part of the scattered light falls into the region of shadows, thereby lighting it up weakly near the edge of the shadow. But the part that is scattered into the region of light interferes with the direct light also arriving in that region. For example, if the detour of the scattered light on its way to point

16. It is easy to observe that the apparent colors change when you look at the oil patch at different angles. If the light penetrates the oil film at an angle, the shift between the two reflected beams is different; hence the color of the nonreflected light is also different.

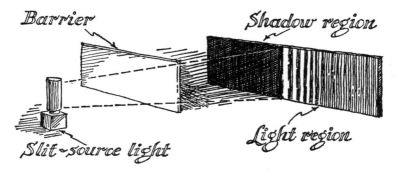

Figure 14
Light and barrier making interference pattern. For the sake of clarity, the pattern of stripes on the screen is shown much wider than it would be with ordinary light.

*A* (see figure 15) is half a wavelength (or three halves, or five halves) longer than the path of the straight light, the two beams will give reduced light. Hence we get a pattern of dark stripes near the edge of the shadow. The smaller the wavelength, the narrower the stripes are. Ordinarily these stripes are invisible to the naked eye, but, as shown in figure 16, they can be observed with instruments.

These phenomena and many others of similar character afford convincing proof that light is a wave motion. They also make it possible to measure the wavelength of light. For example, the thickness of an oil film that does not reflect red light would give us an indication of the wavelength of red light. Such measurements have shown that the wavelengths of visible light are between $4 \times 10^{-5}$ centimeters and $8 \times 10^{-5}$ centimeters, red light having the longer and violet light the shorter wavelength. Since we know the speed of light, we also know how many wave crests per second pass a given point when it rushes by. This number is called the frequency of the

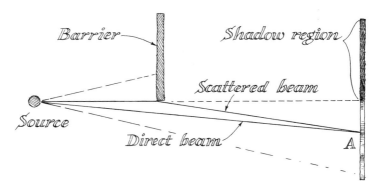

Figure 15
The interference of light. When the direct beam is shorter by $^1/_2$, $^3/_2$, $^5/_2$ . . . wavelengths than the scattered beam (source → edge of barrier →$A$), there will be darkness at $A$. This is the same arrangement as in figure 14, but seen from above. The distance from $A$ to shadow is shown much larger than it would be with ordinary light.

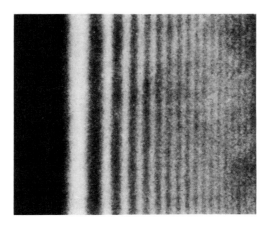

Figure 16
An actual photograph of the interference pattern created by light passing a sharp-edged screen. (From Valasek, *Introduction to Theoretical and Experimental Optics*. New York: John Wiley & Sons, 1949.)

light. It gives the number of vibrations in a light wave per second. Red light has a frequency of $4 \times 10^{14}$ per second, violet light about $8 \times 10^{14}$ per second. This is a tremendously fast vibration that cannot be directly observed.

After we have recognized the wave nature of light, we must face an important question: What kind of waves are light waves? What is the carrier, and what are the oscillatory changes that make up the waves? The answer to this essential question was discovered late in the nineteenth century by James Clerk Maxwell and Heinrich Hertz. The sequence of ideas and discoveries that led to this answer is one of the most exciting developments in science. But before we can give the answer, we must introduce two fundamental phenomena, electricity and magnetism.

## Electricity

A superficial look at natural phenomena does not reveal the pervading importance of electricity. The only obviously electric phenomena in nature seem to be lightning and friction electricity. While the former impresses us with its grandeur and destructiveness, the latter is not impressive at all. Friction electricity is sometimes seen when nonmetallic substances, after being rubbed with some material, attract small pieces of paper or dust and give rise to little electric discharges when brought in contact with metal. These phenomena do not give the impression of being as fundamental as gravity or light and, until the end of the eighteenth century, were considered to be unimportant side phenomena. Nowadays, of course, the importance of electricity is well emphasized in its technical applications; however, the true significance of electricity in nature came to the fore only lately, in the development of atomic physics, when it turned out that almost all the phenomena we see around us in nature are based upon electric forces and their effects.

The first fact to notice is the existence of two kinds of electricity. An object can be charged with either one or the other kind. The two kinds are called positive and negative electricity, but no qualitative distinction is implied in these names. Positive electricity is not "better" than negative electricity. The people who named them could just as well have called the positive kind negative and vice versa. Charged objects exert a force upon each other. If they are charged with the same kind of electricity, they repel one another; if they are oppositely charged, they attract.

The electric charges of opposite kind can cancel each other. A positively charged object can be made electrically neutral if we transfer negative electricity of the same amount to it. Hence if an object is uncharged, it might contain either no electric charge at all, or an equal amount of positive charge and negative charge. It was one of the great discoveries of physics that uncharged matter actually does consist of a combination of positive and negative electricity.

Electric charges can move within matter. The motion of charge is particularly easy within metals. A metal wire connected to oppositely charged objects immediately discharges them, since opposite charges attract each other. The negative electricity of one object is drawn toward the positive charge of the other and vice versa. When a charge moves in a metal wire, we call it an electric current. Nowadays we have readymade "charged objects" in our houses. The two ends in the electric outlets are kept constantly charged with opposite electricity, so that any wire between them gives us a current driven by the electric force between the outlet ends.

Careful investigation of what is moving in a wire has revealed that it is really the negative electricity that moves; the positive electricity stays with the object. The negative electricity consists of small bits of electricity, the electrons, particles that we shall deal with in very much greater length in this book. All substances seem to be filled with electrons.

The negative charge of the electrons in matter normally is balanced by an equal amount of positive electricity. Later we shall see that the positive charge is at the center of the atoms and therefore must remain with the atoms. Electrons can easily be removed from or added to any substance. If some electrons are added, the substance appears negatively charged; if some are removed, there is surplus positive electricity and the substance appears positively charged.

Here we get the first insight into the electrical nature of matter. Superficially matter does not show its electricity; it is hidden by the fact that positive and negative charges in matter normally are balanced exactly, and we do not observe any overall electric charge. Nevertheless, detailed evidence shows that matter actually is made up of electrically charged particles, the movable negative electrons and the centers of the atoms, which carry the positive charge.

Let us return to the force between charged objects. It depends on the distance between the charges. For example, the force between the opposite charges at the terminals of an ordinary electric outlet is too weak to drive electrons from one terminal to the other. If we bring the two ends close enough (about $1/100$ inch), the force will be large enough to get electrons across, and we see a spark.

We can easily measure the force between two charged objects. The attractive force between a positively charged and a negatively charged particle decreases in proportion to the inverse square of the distance, somewhat as the gravitational attraction between two masses decreases with distance. Of course gravity acts between any two masses, whereas the electrical attraction acts only between two objects that are oppositely charged. When it acts between small charged objects, electrical attraction is usually considerably greater than the gravitational force, but it has very much the same character. This analogy brings us to a very fundamental point: The negative electrons in matter are attracted by the positive centers of the atoms in very much the same way the planets are attracted by the sun. Hence we expect the electrons to circle around the atomic nucleus as the planets circle around the sun. This is a conclusion of great importance in the theory of the atom, as we shall see in the next chapter.

**Magnetism**
We are less often sensitive to the phenomenon of magnetism in nature than we are to visible electric phenomena. Only a few materials directly exhibit magnetic properties; some of them are very common—iron, for example. Nevertheless, magnetism is a striking phenomenon; when we hold a magnet and a piece of iron in our hands, we recognize that we are experiencing a special kind of force—a "force of nature" like gravity.

It was a great step forward when it became clear that magnetism is intimately connected with electricity. The connection between magnetic and electric phenomena was discovered by a Dane, Hans Christian Oersted, at the beginning of the nineteenth century. He found that an electric current in a circular or a spiral wire acts exactly like a magnet and creates a magnetic force. This discovery led André Ampère, a Frenchman, to the hypothesis that an ordinary steel magnet must work by the same principle, and he concluded that each iron atom must contain a small circular current. In unmagnetized iron these currents are not oriented uniformly, but one runs in one sense, the next in another sense; therefore, the magnetic effects of the atoms cancel each other. In magnetized iron, however, most of the atomic currents run in the same sense and sum up to produce a net magnetic effect. Ampère's hypothesis has turned out to be entirely correct.

The connection between electricity and magnetism works in both directions. Not only does electricity create magnetism, but magnetism creates electricity. If any magnet is moved in the neighborhood of a metal wire or the wire is moved in the neighborhood of the magnet, a current is produced in the wire. The changing magnetic force has induced a current and, thus, has acted exactly like an electric force. This principle operates in our generators, the machines that produce the electric current we use. There, coils of electric wire mounted on wheels are made to move through magnetic fields when the wheels rotate; electric currents are produced in the wire. Whenever a magnetic field changes, it produces an electric force that sets electric charges into motion.

### Electric and Magnetic Fields

The study of the interactions of electric and magnetic effects led to the discovery of other phenomena in nature, the electric and magnetic fields. This came about in the middle of the nineteenth century, and the most important names associated with it are Michael Faraday, Maxwell, and Hertz. These new ideas have not only deeply influenced our ideas of nature, they have revolutionized our way of life, since they brought with them the development of electrical power and radio transmission. The concept of the electric and magnetic fields is connected with the strange fact that electric charges or magnets exert forces on other objects (charges or magnets) that are not in their immediate neighborhoods. The electric and magnetic forces act through space at a distance. How can this be? What transmits the effects from one body to the next? We meet the same problem also with the force of gravity between two masses.

In order to explain this action occurring at a distance, the physicists have introduced a new concept, the "field." Every object is surrounded by a gravitational field, and every electrically charged object is the center of an electric field. In what follows we will deal with electric (and magnetic) fields and leave gravitational fields alone.

The field is a property of empty space itself. The space in the neighborhood of the charge is in a state of tension. The tension can be measured by means of another charge, a test charge, which will experience a force wherever there is a field to exert one. Hence the attraction of a positive charge $A$ and a negative charge $B$ can be described as follows (see figure 17): Charge $A$ creates an electric field in the space around it. When charge $B$ is placed in this field, it experiences the effect of this field in the form of a force that pushes it toward $A$. In the same way, of course, $A$ is pushed toward $B$ by the field of $B$.

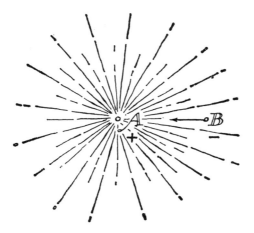

Figure 17
Electric field around a positive charge $A$. A state of tension is created in the space around $A$. The negative charge $B$ is forced toward $A$.

Magnets create a similar field in the space surrounding them, the magnetic field. This is a different kind of "tension" in space. It acts on any piece of iron that is in this region of space; the "tension" takes the form of a force that pushes the iron toward the magnet. (See figure 18.)

Let us now look at the connection between electricity and magnetism and express it in terms of fields. Take the production of a current in a wire by moving a magnet near the wire. When the magnet is moved, its magnetic field changes at the place of the wire: When the magnet comes nearer, the field increases; when it moves away, the field decreases. These changes induce a current in a wire; they bring charges into motion. Hence a changing magnetic field does what an electric field is supposed to do—a changing magnetic field creates an electric field.

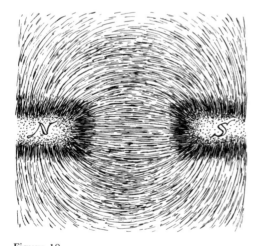

Figure 18
Magnetic field between two poles of a magnet. Iron filings arrange themselves along the field lines as shown.

The reverse is also true: A changing electric field creates in its neighborhood a magnetic field. Let us see how it works. Remember the production of a magnetic field by electricity. Here a current produces a magnetic field. A current consists of moving charges, each of which carries an electric field. Thus we see that moving electric fields create magnetic fields, just as moving magnetic fields create electric ones.

## Electromagnetic Waves

So far the field concept seems to be only a complicated way to describe the forces between charges and the forces between magnets. However, the interaction of electric and magnetic phenomena shows that these fields actually have an existence of their own.

The relationship between electric and magnetic fields occupied physicists' minds through the middle of the nineteenth century. The man who contributed most to this problem, and who was able to bring mathematical order into this topic, was the great English theoretical physicist James Clerk Maxwell. The mathematical relations that connect the two fields and are the basis of all our knowledge of electric phenomena are called the Maxwell equations. Their conception was a turning point in our understanding of nature, and it led to innumerable developments in physics and technology, of which the world of radio, radar, and television are only a few of the many.

Maxwell studied the relationship between the two fields in detail and arrived at the following interesting question: If electric and magnetic fields are entities in themselves, shouldn't they exist independently of charges and magnets? Certainly static (unchanging) fields can exist only around charges and magnets, but what about changing fields? The fact that changing fields produce fields of the other kind suggests the possibility that this process might be self-perpetuating. A changing electric field creates a magnetic one; the latter field changes its strength in the process of being created; hence it again creates a new electric field, and so forth. From the quantitative analysis of these relationships, Maxwell could show simply that this process propagates in space—a varying electric field at one point produces a magnetic field in its neighborhood, which in turn produces an electric field a little farther off, and so on. Therefore, we obtain an electromagnetic field constantly expanding into space, an electromagnetic wave. Whenever a changing electric or magnetic field is produced—for example, by oscillating charges or magnets—the field will propagate in all directions. The speed of this propagation can be figured out from the observed strength of the currents induced by moving magnets and from the observed strength of the magnetic fields produced by currents. One finds that the speed of propagation of electromagnetic fields must be $3 \times 10^8$ meters/sec, and this is exactly the speed of *light*.

## What Is Light?

It was one of the great moments in the history of science when Maxwell completed those calculations. He used only the measurements of electric currents and magnetic fields, phenomena that have seemingly nothing to do with light, and he concluded from these measurements that oscillating electric fields propagate like waves through space, exactly with the velocity of light signals. A connection was discovered between two parts of physics that appeared to be totally unconnected: optics and electricity.

From Maxwell's calculations it was then only a small step, but still a bold one, to conclude that light is nothing but a propagation of electromagnetic fields. With this recognition many scattered facts fall into order. For example, we understand immediately why all matter emits light when heated to a high temperature. It comes from the electric composition of matter. At high temperatures the charged particles of matter, in particular the electrons, perform intense and rapid motions; therefore, they create rapidly changing electric fields that initiate the propagation of fields into space at the velocity of light. Light is emitted.

If Maxwell's idea of the electromagnetic nature of light is correct, one should be able to produce new kinds of light. Any electric charge or magnet set into oscillation would generate electromagnetic fields propagating into space and act as a source of light at a frequency equal to the frequency of the oscillations.

For example, an oscillating current in a wire would send out electromagnetic waves, and one might pick them up at a large distance if he exposed another wire to the waves and observed the weak currents induced in it. This experiment was first done in 1880 by Hertz, who wanted to prove the correctness of Maxwell's ideas. Hertz's success opened up a whole new technology. Today space is full of such waves. They are the radio waves, which are produced by alternating currents in antennas and differ from ordinary light waves only in their frequency and wavelength. Evidently the current oscillations that we set up artificially in our antennas are much slower than the oscillations of electrons heated in a light bulb. So the radio waves are light, but with much lower frequency and consequently much longer wavelength. (See figure 19.)

Now we also can answer the question about the nature of light waves: What is it that oscillates and what carries the wave? It is the electric- and magnetic-field strength that performs the oscillations. The carrier of the wave is space itself,

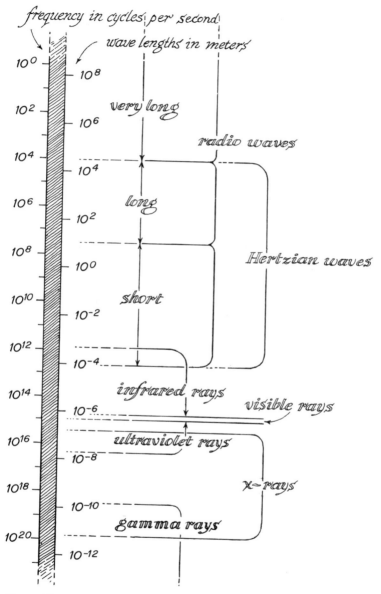

Figure 19
Electromagnetic spectrum.

since the electromagnetic fields are properties of space; they are space under tension. The electric and magnetic tensions travel through space as light, just as the compression and dilution of air travels through air as sound. An electromagnetic wave is a double wave: electric- and magnetic-field strengths travel together and are intimately interwoven. Thus light is electromagnetic. It is electromagnetism in its purest form: Light is a wave of electromagnetic fields traveling through space, separated from the electric charges that have produced them.

Maxwell's discovery is comparable in its importance to Newton's discovery of the law of gravity. Newton connected the phenomenon of planetary motion with the phenomenon of gravity on Earth and discovered the fundamental laws governing the mechanical motion of masses under the influence of forces. Maxwell connected optics with electricity and discovered the fundamental laws (the Maxwell equations) that govern the behavior of electric and magnetic fields and their interaction with charges and magnets. With Newton's work the concept of a universal law of gravity was introduced; Maxwell's work established the concept of the electromagnetic field and its propagation in space.

# 4 Atoms

## The Natural Units of Matter

In our world we find an overwhelming variety of different substances with most complicated structures and properties, in particular when we look at living matter. In order to get at the fundamental features of the structure of matter, we must begin our study with simple substances. At the start we shall not consider organic substances, such as wood or the skin of our bodies, whose structure is intricate and seems to be a complicated combination of substructures. We shall first consider homogeneous substances such as air, water, oil, a piece of metal, or a sample of rock. These substances occur in three states of aggregation—in the solid state, the liquid state, and the gaseous state as vapors. In the solid and liquid states matter seems to be densely packed; it is extremely difficult to compress matter in these states. In the gaseous state compression is very easy; therefore, one would conclude that matter in a gas is dilute, that there is empty space between the units of matter.

What are those units of matter? Do units indeed exist? Can we subdivide a certain amount of a given substance indefinitely or is there a smallest amount? The answer to this fundamental question is well known today. There is a smallest unit of every substance and it is called a molecule, and in some substances an atom. The difference between an atom and a molecule will be taken up in the last section of this chapter. Until then we do not need to distinguish between those two types of smallest units. The units are very small, and most of the simple substances give the impression of homogeneity. However, investigations with very fine instruments reveal that there is a molecular structure. Figure 20, for example, shows a picture of a tip of a very fine tungsten needle taken with a so-called field-ion microscope, a device with which one can locate extremely small details on some metallic surfaces. Here

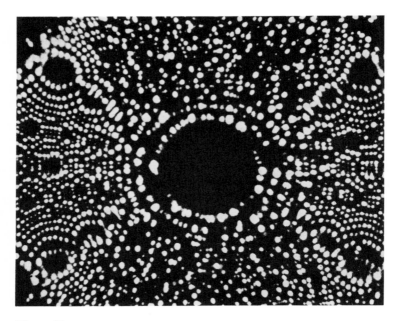

Figure 20
This remarkable photograph of the arrangement of tungsten atoms was made with a field-ion microscope. (Courtesy of Erwin W. Müller and Paul Weller.)

we see a regular structure of the units of which tungsten seems to consist. We can calculate from the magnification of this microscope how small the units are; we find that they measure roughly $3 \times 10^{-8}$ centimeters. A 100 million times this length is about one inch. We can use our experience from the first chapter to appreciate this small size. The breadth of a hair is 10,000 times smaller than the length of an outstretched arm (one meter). But 100 million is 10,000 multiplied by 10,000. Hence the size of a tungsten atom is to the size of one inch as the breadth of a hair is to ten kilometers (six miles).

The molecular nature of a gas, such as air, can be shown in a very impressive way. We know that air can move light objects; it is moving air that sets the leaves of a tree rustling. But if air is kept at rest in a container with no wind or current, we do not expect to detect any movement of objects that are suspended in the still air. If the objects are very small and light, perhaps very small particles of dust or smoke, our expectation is in for a shock. If you look through a microscope at particles suspended in air, you will notice that they undergo small irregular displacements here or there. (See figure 21.) It looks as if they were being hit with invisible tiny bullets fired at them at random from all directions. This irregular motion of small particles was first discovered by a botanist, Robert Brown, in 1827, when through his microscope he found small particles performing this random dance. The particles he observed were immersed in water and not in air, but the principle involved is the same.

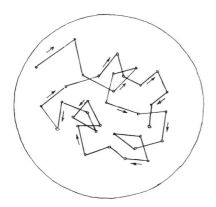

Figure 21
Brownian motion. A lightweight particle suffers irregular displacements in air if seen through a microscope.

This "Brownian motion" is direct evidence that air is not continuous, but consists of many small units flying through space in all directions at random and in an irregular pattern. Any object in air is bombarded at random from all sides by the molecules, and this bombardment produces the air pressure. Ordinarily the number of hits is so large that it causes a continuous pressure effect on all sides equally. If the object is very small, however, there are far fewer hits altogether; then, at one instant, more molecules will hit one side than the opposite side, causing an unbalanced force. At a later instant the unbalanced force will act in another direction. This is the cause of the Brownian motion.

We can learn about the smallest units of liquids by asking a question: How large an area can we cover with a thin film of liquid when we have only a given quantity of liquid at our disposition? If there were no smallest unit, one could cover any area with a film of one milligram of liquid, since one could always double the area by making the film half as thick. But if a smallest unit does exist, the thickness of the film cannot be less than the size of this unit, and there must be a largest area over which a given quantity of liquid can spread.

This experiment can be performed very easily by spreading a quantity of oil on a water surface. It turns out that a little droplet of oil, one millimeter across (approximately a milligram) spreads over as much as thirty square feet, but never farther. We conclude, therefore, that there is a smallest unit of oil. We can calculate from the size of the area how large the unit is, and we get roughly $3 \times 10^{-8}$ centimeters for the size of the oil unit, not very different from the tungsten atom.[17]

17. Let us assume that the drop of oil one millimeter across has the shape of a cube one millimeter high. Let us then reassemble the cube by cutting the thirty square feet of oil sheet into small squares of one millimeter size and piling them up on top of each other. The pile will reach a height of one millimeter—the height of the original cube. Thirty square feet contain three

The existence of a definite smallest unit in every substance supplies us with an absolute measure of quantity. Ordinarily we speak of one pound of iron, one gallon of water, a cubic foot of air under atmospheric pressure. All these measures depend on arbitrary conventions of measurement. But when we refer to one million iron atoms, one million water or air molecules, then we have made use of an absolute measure that is characteristic of the substance and is independent of any human convention. Matter can be "counted" instead of weighed or measured.

Since the molecules or atoms are such extremely tiny quantities from our human point of view, the chemists prefer to use as an absolute measure of matter, a "mole" of a substance. This is a fixed number of smallest units; for practical reasons, the number of atoms in one gram of hydrogen has been chosen to define the mole. It is the famous Avogadro's number.

## Avogadro's Number    $6.03 \times 10^{23}$

One mole of water, which is $6.03 \times 10^{23}$ water molecules, fills a little more than a cubic inch; one mole of rock (quartz) fills about $4/3$ cubic inches, and one mole of air under normal conditions is contained in $3/4$ of a cubic foot. The mole of water and the mole of quartz fill roughly the same space; so the size of their smallest units must be about equal. The mole of air, however, fills a much larger volume than the mole of water and rock. This is not because of the large size of the unit; it is because air is a gas, with its molecules far apart from each other, flying freely around in space. The atoms in air are ten times farther apart than in a solid. This means a volume

---

million square millimeters. Hence the thickness of the oil sheet is one millimeter divided by three million. This gives about $3 \times 10^{-8}$ centimeters, which is then the size of the oil unit, the molecule of oil.

greater by a factor $10 \times 10 \times 10 = 1,000$, which accounts for the immense differences between solids and gases in volume as well as in other properties. When we cool air to such a low temperature that it becomes a liquid (the molecules then touch each other), the volume of a mole is roughly the same as for water.

## Heat and Random Motion

What makes an object hot or cold? For a long time people believed that heat was a substance contained in a hot object. When in contact with a cold object, the heat substance, they said, diffused into the cold object and thus equalized the temperature. In the middle of the last century it became clear, however, that heat is energy, specifically the energy of random motion of the molecules or atoms. When a sample of matter is heated, all that changes is that the smallest units perform faster and more energetic random motions.

Let us look at a few examples. In the tungsten needle we saw the tungsten atoms arranged in a regular pattern. What are the motions they perform when heated? They vibrate and oscillate around their assigned places in the regular structure. Indeed this back-and-forth motion is partially responsible for the fuzziness of the picture in figure 20. At higher temperatures the picture would become even fuzzier. If we raise the temperature very much, the vibrations become great enough to destroy the regular arrangement. This happens when there is enough heat to melt the metal.

In a gas such as air the heat motion of the molecules is the ordinary straight motion of each molecule when the molecules fly at random through space, colliding with each other and with the walls. The higher the temperature, the faster the motion. On a cold day ($-20°$ celsius) the average speed of an air molecule is around 400 meters per second (900 miles per hour); on a hot day (30° celsius) it is about 440 meters per second (1,000 miles per hour). The difference between a very hot and a very cold day is only a 10 percent difference in speed of the air molecules. We notice this difference of molecular speed in the pressure of automobile tires. The air pressure is caused by the impulse of the molecules colliding with the walls, and the effect of this impulse is proportional to their speed. A decrease in speed of 10 percent reduces this impulse by 10 percent. In addition, a 10 percent drop in speed means 10 percent fewer impacts per second. Hence when the temperature falls from (30° to −20° celsius), the tire pressure drops by 20 percent. In spite of the tremendous speed of air molecules at ordinary temperatures, the molecules don't travel very far in a straight line. Collisions with other molecules constantly interrupt their flight. In air their average free and uninterrupted flying path is only one hundred thousandth of a centimeter. Hence their flight is more like a fast random motion at the previously mentioned speed, but with a change of direction after every hundred-thousandth centimeter.

When the temperature falls, the average speed decreases. If one goes on lowering the temperature, one reaches a point at which heat motion ceases completely. Indeed, heat motion in any material, solid or gas, ceases at a temperature of −273° celsius, which is also called zero-degree absolute temperature or zero-degree Kelvin. This is the temperature at which all molecular random motion is frozen still, and it is obviously the lowest possible temperature.

## Molecules and Atoms

Can the smallest units of matter be broken into even smaller parts? They can, but the parts are no longer the same substance. A molecule of water is the smallest unit of water. Part of the molecule is no longer water, it is hydrogen or oxygen.

It is much harder to break a molecule into smaller parts than to break a substance into its molecules. For example, when we boil water and produce steam, we have broken up the substance water into its molecules. In the form of steam, water is a gas whose smallest units fly about in space, each molecule separated from the other, but each molecule itself remains an unbroken unit. When we discharge a strong electric spark through steam, however, some of the molecules will be broken and we get hydrogen and oxygen gas. The spark is a much stronger energy source than the process of boiling. In some cases intense heating breaks up the molecules and thus changes one substance into another.

In the development of chemistry through the ages, we have come to recognize that certain substances can be decomposed into others by intense heat, electric sparks, or other violent treatments, and that two substances can be compounded to form a new substance, a chemical compound. Hydrogen and oxygen can be compounded into water, and a piece of quartz can be decomposed into silicon and oxygen.

One of the most exciting moments in the history of mankind must have been the time about 3000 B.C. when a man first put certain earthlike substances, perhaps cuprite or galena, on a charcoal fire. (See figure 22.) Out came a new substance, metallic copper or lead. Most metals such as iron, copper, lead, zinc, etc., are truly man-made substances; they rarely occur in nature except for the very small amounts of

Figure 22
Woodcut illustration from sixteenth-century book on metallurgy depicting blast furnaces for the smelting of copper ore. The foreman is tapping the furnace with a hooked bar, letting the melt flow into the forehearth. Below is a dipping pot with a tap hole to the forehearth. The assistant carries charcoal from the pile, left. (Reprinted from Bern Dibner, *Agricola on Metal*. Norwalk, Conn.: Burndy Library, Inc., 1958.)

native ores (copper, for instance) and nickel-iron alloys that come from outer space in meteorites. There is a simple reason: Pure metals do not last when exposed to the oxygen in air. Most metals combine with oxygen in time and form chemical compounds, which are the same earthlike substances from which they were extracted. Man can transform these ores into pure metals, but they remain pure metals for only very short periods compared to the age of the earth.

The study of the processes in which substances are changed into other substances has revealed a significant fact: All, really all, existing substances can be decomposed into ninety-two fundamental substances, which are called the elements. Any piece of matter, wherever found and in whatever state of aggregation, is always either an element or consists of several elements. A substance whose smallest unit is a combination of several elements is called a chemical compound. Many familiar substances are actually elements. Metals such as gold, silver, iron, lead, aluminum, etc., are elements. Many gases, such as hydrogen, oxygen, or nitrogen, are also elements, but other gases, such as illuminating gas or carbon dioxide, are compounds. Most well-known liquids are chemical compounds. The smallest unit of an element is called an atom. The smallest unit of a chemical compound is called a molecule. Since all chemical compounds can be decomposed into elements, the smallest unit of a chemical compound must be made up of the smallest units of elements. Hence each molecule is a conglomeration of atoms; it consists of the atoms of those elements that make up the chemical compound. They fit together and form a stable unit, the molecule, which is endowed with all the properties of the substance of which it is a unit.

Water is a chemical compound of hydrogen and oxygen. The smallest units of the elements hydrogen and oxygen are the hydrogen atoms and the oxygen atoms. The smallest unit of water is the water molecule, which consists of two hydrogen atoms and one oxygen atom ($H_2O$), so tightly bound together that only an electric spark can break them apart.[18]

There are small and large molecules. The molecule of water consists of only three atoms; a molecule of ethyl alcohol consists of nine atoms: one oxygen, two carbon, and six hydrogen atoms. Some molecules occurring in living matter, such as proteins, contain hundreds of thousands of atoms.

The discovery of the ninety-two elements and their atoms was the most important step toward the understanding of the structure of matter. It took a long time for the ideas to clear and the facts to be recognized. The concept of basic substances of which all other substances can be made is as old as natural philosophy. Many Greek philosophers speculated with ideas of this kind. The first conclusions similar to our present ones were drawn by Robert Boyle in the seventeenth century, although many of the substances that he thought were elements turned out to be chemical compounds. The famous French chemist Antoine Lavoisier, who was killed in the French Revolution, made a list of thirty-three elements. These thirty-three compose more than 99 percent of the matter in the universe. The present list of elements and the interpretation of molecules as being combinations of atoms of elements were developed in the nineteenth century, and the most important work came from the English chemist John Dalton.

18. There is one fact that often causes confusion: Many elements occur naturally in the form of molecules consisting of a pair of atoms of the element. This is true of most of the elements appearing as gases—hydrogen, oxygen, etc. ($H_2$, $O_2$, etc.). If such a molecule is broken, the process does not change the substance. For the sake of clarity, however, we always consider the atom as the smallest unit of an element.

Let us be aware of the immense impact of this discovery. We are surrounded by an infinite variety of substances with different and ever changing forms, shapes, and qualities—hot or cold, living or dead. In spite of this immense variety, every object we know about is made of only ninety-two different kinds of atoms, each kind belonging to a well-defined, specific element. Nothing is ever found in living or nonliving matter that cannot be decomposed into some of the ninety-two elements. This discovery revealed a basic simplicity in the structure of matter. We have to deal with combinations of a relatively small number of fundamental units. Hence there is some hope that the principles underlying the structure of matter are simple enough to be grasped by the human mind.

**The Internal Structure of Atoms**
What is the structure of the atoms themselves? What is it that exists in ninety-two different forms and is endowed with an ability to combine and produce the great variety of known substances? We must be able to understand why certain combinations are possible and others are not; and finally, we must try to get some understanding of how such highly organized systems as living matter can originate.

The ninety-two kinds of atoms have very different properties. Some form gases; some form metals. Some, such as the carbon atoms, are able to combine easily with other atoms and form the backbone of many chemical compounds. Others, such as the atoms of helium, neon, or argon almost never combine. In spite of these differences, the atoms all seem to be roughly of equal size.

When we know the atomic constitution of the molecule of a substance, we can find out easily how many atoms are contained in a given amount of that substance. Let us remember that a mole of water is that quantity containing $6.03 \times 10^{23}$

molecules, and that this quantity fills a little more than a cubic inch. Since a molecule of water contains three atoms (two hydrogen and one oxygen), the mole of water contains about $18 \times 10^{23}$ atoms; so one cubic inch of water then contains a little less than this number of atoms. We get a similar but smaller value for the number of atoms in a cubic inch of rock: one mole of quartz fills $4/3$ cubic inches. One molecule of quartz consists also of three atoms, one silicon and two oxygen. Hence the number of atoms in a cubic inch is $3/4 \times 18 \times 10^{23} = 13.5 \times 10^{23}$ atoms. Even when taking very different substances, liquids, or solids such as gold or wood or carbon, one always gets something between 10 and $25 \times 10^{23}$ for the number of atoms per cubic inch. Since the molecules are closely packed in liquids and solids, and the atoms are also closely packed within the molecules, we conclude that all atoms are roughly of the same size; about 10 to $25 \times 10^{23}$ per cubic inch. This means that the size of an atom is a few $10^{-8}$ centimeters across, as we saw earlier for the example of tungsten.

What do we know about the internal structure of atoms? Here we come to a fundamental question—the mechanisms that we find in the atoms must be the clue to an explanation of the properties of the matter we see around us. It became clear in the last chapter that electricity plays an important role, and that the electrons are an essential part of the atom. The decisive experiments, however, were made in 1911 by Ernest Rutherford, Hans Geiger, and E. Marsden, who probed the structure of the atom with alpha particles, the very fast electrically charged particles emitted by some radioactive substances. They directed a beam of these particles through a metal foil and observed where and how much the direction of motion of the particles changed when penetrating the metal. (See figure 23). These measurements tell us something about how the electricity is distributed in the atoms of the metal. If

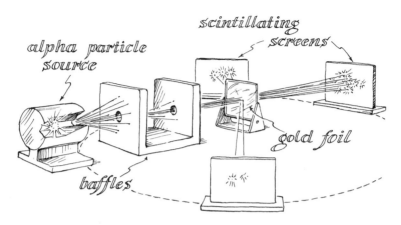

Figure 23
Rutherford experiment.

electric charge is spread smoothly over the atom, the alpha particle on its flight through the atom would experience smooth electric forces and would never be strongly deviated from its path. If electricity is concentrated at certain points in the atom, however, the alpha particle would be strongly influenced whenever it came near these points. It is worthwhile to quote Rutherford's own description of this experiment:

. . . I would like to use this example to show how you often stumble upon facts by accident. In the early days I had observed the scattering of α-particles, and Dr. Geiger in my laboratory had examined it in detail. He found, in thin pieces of heavy metal that the scattering was usually small, of the order of one degree. One day Geiger came to me and said, "Don't you think that young Marsden, whom I am training in radioactive methods, ought to begin a small research?" Now I had thought that too, so I said, "Why not let him see if any α-particles can be scattered through a large angle?" I may tell you in confidence that I did not believe that they would be, since we knew that the α-particle was a very fast, massive particle, with a great deal of energy, and you could show that if the scattering was due to the accumulated effect of a number of small scatterings the chance of an α-particle's being scattered backwards was very small. Then I remember two or three days later Geiger coming to me in great excitement and saying, "We have been able to get some of the α-particles coming backwards. . . ." It was quite the most incredible event that has happened to me in my life. It was almost as incredible as if you fired a 15-inch shell at a piece of tissue paper and it came back and hit you. On consideration, I realized that this scattering backwards must be the result of a single collision, and when I made calculations I saw that it was impossible to get anything of that order of magnitude unless you took a system in which the greater part of the mass of the atom was concentrated in a minute nucleus. It was then that I had the idea of an atom with a minute nucleus. It was then that I had the idea of an atom with a minute massive center carrying a charge. I worked out mathematically what laws the scattering should obey, and I found that the number of particles scattered through a given angle should be proportional to the thickness of the scattering foil, the square of the nuclear charge, and inversely proportional to the fourth power of the velocity. These deductions were later verified by Geiger and Marsden in a series of beautiful experiments."[19]

With these experiments and many others that followed, it was shown beyond doubt that the atom consists of a positively charged, small but massive nucleus, in which most of the weight of the atom resides, surrounded by negatively charged electrons, which are much lighter than the nucleus. The actual size of the nucleus is extremely small. Its diameter is between $10^{-13}$ and $10^{-12}$ centimeters, depending on the kind of atom; it is therefore about 10,000 times smaller than the atom itself, but quite heavy since it contains almost all the

19. E. Rutherford, "The Development of the Theory of Atomic Structure," *Background to Modern Science* (New York: Macmillan, 1940).

mass. Rutherford and other physicists, in particular Moseley, determined the number of electrons in each atom and the charge of the atomic nucleus. Since the total atom is uncharged, the negatively charged electrons must balance the charge of the positively charged nucleus. Hence the number of electrons must always be equal to the charge of the nucleus expressed in units of electronic charges. This number is characteristic of each kind of atom. Hydrogen, for example, has one electron and one positive-charge unit in the nucleus; helium has two electrons, lithium three, etc., up to uranium with ninety-two electrons and a nucleus charged with ninety-two positive-charge units. This characteristic number is called the atomic number Z. Every element has its characteristic atomic number Z, which gives the positive charge of the nucleus and also the number of electrons in the atom.

With this discovery the qualitative difference between the ninety-two elements was reduced to a quantitative one: Atoms of one element differ from atoms of another by the number of electrons, a number that also determines how many positive-charge units there are in the nucleus.

One can order the atoms according to the atomic number Z, and each number from 1 to 92, except technetium (43) and promethium (61), corresponds to an element found in nature. Here is the number Z for the most important elements:

| | | | |
|---|---|---|---|
| Hydrogen | 1 | Silicon | 14 |
| Helium | 2 | Iron | 26 |
| Lithium | 3 | Silver | 47 |
| Carbon | 6 | Gold | 79 |
| Nitrogen | 7 | Lead | 82 |
| Oxygen | 8 | Uranium | 92 |
| Sodium | 11 | | |

There also exist artificially produced elements, the so-called "transuranic" elements, which have more than ninety-two electrons. They have short lifetimes and do not occur in nature under normal conditions.

## The Great Problems of Atomic Structure

This reduction of the qualitative differences among ninety-two kinds of atoms to a quantitative one represented an enormous step forward. But every great scientific discovery creates new problems even as it solves old ones. When we know more, we have more questions to ask. Our knowledge is an island in the infinite ocean of the unknown, and the larger this island grows, the more extended are its boundaries toward the unknown. The recognition of structure in the atom immediately poses a question: How can those quantitative differences of structure be the cause of the observed qualitative differences in the properties of the elements? How is it possible, for example, that bromine, with thirty-five electrons, is a brownish liquid forming many characteristic chemical compounds, while krypton, with thirty-six electrons, is a gas forming no compound at all, and rubidium, with thirty-seven electrons, is a metal? Why should one electron more or one electron less make such a great difference in the properties of the atom? This question was not answered until later, when the quantum nature of matter was understood. It will be the subject of the next chapter.

What motions do we expect in the atom? When Rutherford found that the atom consists of a massive positive center surrounded by light negative electrons, it seemed likely that the atom was similar to a planetary system. The electrons are attracted to the center by the electric attraction between opposite charges. This force is much stronger than the force of gravity, but it obeys the same law in its dependence on the distance—it decreases with the square of the distance. Therefore one may expect the electrons to move around the nucleus in much the same way the planets move around the sun.

Electric attraction between nucleus and electron replaces the force of gravity. According to this picture, an atom should be a small planetary system, and each kind of atom should have a different number of planet electrons. We might expect to find in the small world of the atom a replica of the big world in the sky.

In some respects this expectation seemed to be satisfied. For example, we can calculate what number of revolutions an electron would make around the nucleus per second, say, in hydrogen. We know the size of the orbit—it is about as big as the hydrogen atom, roughly $10^{-8}$ centimeters. We then know the force with which the electron is attracted, and we can figure out the speed at which it must circle the nucleus in order to make the centrifugal force equal to the attractive force. This calculation gives about $10^{16}$ revolutions per second, which means that the "year" of the atomic solar system—the time the planet takes to get around once—is as short as $10^{-16}$ seconds. This number can be put to a test. We know that oscillating electric charges radiate light and that the frequency of this light (the number of ups and downs of the light wave per second) must be equal to the number of the oscillations per second of the electric charge. Therefore we should expect that the light emitted by a hydrogen atom has a frequency of $10^{16}$ per second. In fact, incandescent hydrogen gas does radiate light of approximately that frequency.

But the planetary atomic model soon runs into great difficulties. If the atom were a true planetary system in which electric charges are constantly circling the nucleus, the revolving electrons should emit light all the time, in ordinary cool hydrogen as well as in hydrogen incandescent under very high temperatures. This does not happen. There is another important shortcoming: The light of hydrogen gas, and of any other gas, too, is emitted and also absorbed only at definite frequencies that are characteristic of the element making up the gas. It is as if every atomic species is a radio station to which have been assigned certain specific frequencies for transmitting and receiving signals. The spectroscopists have studied these assignments over many decades. The frequencies provide an excellent tool for the identification of elements; it is as if one would identify a radio station by looking up its frequency in the lists of radio transmitters. It is the only way to get information about the chemical compositions of the stars.

Now it is very difficult to reconcile this situation with a planetary system structure. There are many possible orbits around the center. In some of these orbits around the nucleus, the electron circles faster than in others. The question arises why the electron should circle only in those orbits that have the assigned frequency. It is all the more incomprehensible since we know that in a gas of atmospheric pressure the atoms collide with each other about $10^{11}$ times a second (that, on the average, is once in 100,000 hydrogen-atom years). The energy of these collisions can be deduced from the heat energy in a gas. The impacts are quite powerful and should change the orbits of the electrons completely in respect to their size, shape, and frequency. How, then, is it possible that they keep their assigned frequencies?

In order to illustrate this question more vividly, let us consider a sample of sodium gas. It absorbs only light that has the particular frequency assigned to the sodium atom. When heated up, sodium emits the well-known yellow sodium light, its assigned frequency. Let us now condense the gas to a solid piece of sodium metal by cooling or compressing. In the metal the atoms touch each other and therefore the planetary orbits intermesh. We shall not be astonished to find that the metal does not respond particularly to the frequency assigned to the free sodium atom. In fact, the metal does not seem to have any particular frequency of response, as one might expect of a complicated intermesh of electron orbits. Let us then transform the metal back into sodium gas by evaporation. The gas will have exactly the same properties as before: it will absorb and emit only the frequencies typical for the sodium atom.

This behavior is utterly at variance with, and completely incomprehensible on the basis of, a planetary atomic model. These are properties one would never expect of a planetary system. How could we imagine that the electrons will find their way into exactly the same orbits when the atoms are evaporated from the metal? There is not the slightest reason for it. In fact, it would appear improbable to the highest degree that the orbits after evaporation will be similar at all to the orbits before condensation, except in general shape and in approximate size. But what we do find is an equality of frequency and of many other features, to a degree that is accurate to the most minute details. It is as if the planet Venus, after having been knocked out of its orbit in some collision with another star, should obediently glide back into its previous orbit when the star has gone.

We are accustomed to finding in nature substances with well-defined and reproducible properties. It is deeply ingrained in our way of thinking that nature is so, and we are not at all astonished that, for example, two atoms of gold, mined at different locations and processed in different ways, end up identical, indistinguishable from one another. All our lives are built on the experience that substances have their characteristic properties; we are able to recognize metals, minerals, and chemicals and distinguish between different kinds of substances on the basis of their characteristic and ever recurring properties. Gold always has the properties of gold, and the seed of a zinnia will produce zinnias every spring.

We must realize, however, that all this remains incomprehensible on the basis of the planetary model of the atom. Not only is it beyond explanation, but it is opposed to the most characteristic features of a planetary system. The structure of the orbits is bound to depend on the initial conditions; there are many possible forms and shapes of the orbits that depend on the previous history of the system. Only very rarely would two atoms of the same kind exhibit identical properties if they were ordinary planetary systems.

Let us summarize the situation: All about us nature exhibits characteristic and specific properties of various materials. In spite of the overwhelming variety of substances, each substance is reproducible and recurrent with all its characteristic properties. For this situation to exist, the atoms must have three properties:

1. *Stability* The atoms keep their specific properties in spite of heavy collisions and other perturbations to which they are subjected.

2. *Identity* All atoms of the same kind (same electron number Z) exhibit identical properties; they emit and absorb the same frequencies, they have exactly the same size, shape, and internal motion.

3. *Regeneration* If an atom is distorted and its electron orbits are forced to change by high pressure or by close neighboring atoms, it regains its exact original shape and orbits when the cause of distortion is removed.

Experiments indicate, however, that the atom is a planetary system of electrons circling around the nucleus, a system that should never exhibit these three properties. Hence this picture of the atoms cannot explain fully the specificity of material qualities. We must find a new and essential trait in the structure of the atom that is not contained in the classical picture of the atom as a planetary system. This new insight into the nature of the atom was provided by the development of quantum theory.

# 5 The Quantum

The world of atoms is full of the unexpected. When we try to penetrate into the inner structure of the atom, we observe strange things that appear contradictory because they are so different from our experience with ordinary large-scale matter. Atoms behave in ways that do not make sense when compared with our descriptions of ordinary particles and our expectations of how they should behave. We are aware that something new and unusual must be found if we are to explain the facts of nature as we see them around us.

In the preceding chapter we stressed the serious contradictions afflicting the study of the structure of the atom. On the one hand the atom seems to be a small planetary system of circulating electrons; on the other hand we find a stability and an exhibition of characteristic properties completely foreign to a planetary system. In this chapter we start out with a more detailed account of further unusual observations concerning atoms and atomic particles and hope to find our way to the new phenomena that govern the interior of atoms. It will not be a historical account. In the actual progress of science, unfortunately, a discovery is rarely made at the time when it could be most useful for our understanding; it does not come until technological development has created the means of performing the necessary measurements. Here we will put the new discoveries into an order that makes it easier to discern the deeper sense in them.

We shall discuss three groups of observations, each one revealing strange and uncommon features of the atomic world. The first contains the discoveries of quantum states in the atom; the second deals with the wave properties of material particles; and the third deals with the quantum nature of light. Then we should be ready to understand the essential content of the new mechanics of the atom, called quantum mechanics, which is based upon these discoveries. It is the framework of our present understanding of atomic phenomena.

## The Quantum States of the Atom

In 1913 James Franck and Gustav Hertz performed a series of experiments in which they attempted to change the planetary orbits of the electrons in the atom. They argued that the atom seemed to resist changes of its electron orbits and they attempted to change these orbits by force and see how, and how much, the atom could resist. We would expect that the orbits of the planets would be changed if a star should pass close to our solar system. Franck and Hertz arranged an experiment in the atomic world that would correspond to such a solar cataclysm. In simple terms, their experiment was this: We have a container filled with a gas of atoms, in their case it was mercury vapor (figure 24). We pass a straight beam of electrons through the gas. Since electrons have a strong electric effect on each other, we expect that an electron of the beam, when passing near an atom, will influence the orbiting electrons in the atom and change their orbits, just as a nearby passing star would change the orbit of the earth.

We cannot look directly at the electron orbits and see whether they have been changed, but we can find out indirectly what has happened. We make sure that in the beam of electrons all electrons have exactly the same speed when they enter the gas. Any change the electrons may cause in the atom will be associated with a change of speed of the electrons. This prediction follows from the law of conservation of energy: When something gains energy, something else must lose an equal amount of energy.[20] Energy is needed to alter the orbit of

20. Whenever something happens in nature, energy is exchanged. If I rap on the table with my finger, energy is transmitted from my body to the table, and the energy of my body is reduced by this amount. I shall have to eat again in order to replenish it.

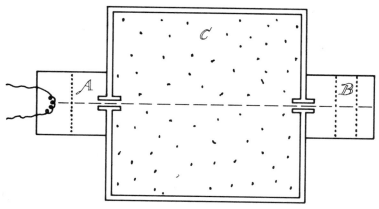

Figure 24
The general idea of an experiment to measure changes in the energy of electrons when they collide with gas atoms. The electrons are produced in the electron gun *A*, where they acquire certain definite energy. They then enter the container *C*, which is filled with the atoms under investigation (for example, sodium atoms). The electrons then pass into the chamber *B*. Some of them have lost a certain amount of energy; the loss is measured in the chamber *B*.

an electron in an atom; hence if the orbit is changed by a beam electron passing by, the electron must lose some energy. Less energy means less speed; therefore the electron's speed will be reduced, and this reduction can be observed when the beam leaves the gas on the other side. The same would happen if a star passed by our solar system. Its passing might give a push to the earth, thus increasing the earth's energy while decreasing the energy of the star.

What should we expect on the basis of the planetary model? There should be all kinds of changes of orbits, small and large, depending on how close the electron has passed by an atom. We should expect energy losses (or sometimes gains) ranging from zero and up; the average loss should be less when the beam goes through a more dilute gas, since there would be fewer close approaches.

The observations turned out to be completely different. No change of speed at all was observed if the initial energy of the electrons was less than a certain minimum. This minimum energy was quite high—more than a hundred times greater than the heat energy of electrons at ordinary temperatures. When the energy was higher than that minimum, the electrons lost either certain specific amounts of energy or none at all. These specific amounts, and also the minimum, are characteristic of the kind of atom in the gas; they do not depend on the density of the gas or on any other external circumstances.

What can this strange result mean? It tells us that one cannot change the electron orbits in the atom by a small amount. Either they do not change at all or they change by specific, and rather large, amounts of energy. Here the concept of the *quantum* of energy comes in. Energy can be fed into an atom only in certain characteristic quanta—no more, no less.

It is as if the atom accepts energy only in predetermined lumps. It does not take a small bite, but only the full lump. Every atom has its own characteristic lumps of energy that it can accept. If less is offered, the atom does not budge at all. In fact it budges (changes in state) only if it is offered just the right amount.

This situation is certainly foreign to our picture of a planetary system. A passing star can feed any amount of energy into the earth's orbit. Typically the greater the distance of passing, the smaller the energy transferred. But the result of our experiment is not so startling in view of what we already know about the atom. It shows that the state of the atom has an intrinsic stability. Weak impacts cannot change it, only a large amount of energy. There must be some reason why the atom stays in its lowest energy state and a change can be induced only by a large energy. Could it be the same reason that accounts for the specificity of the atoms, which always forces the electrons back into the configuration characteristic of the special type of atom?

We must be more quantitative now. What is the minimum energy necessary to change the state of an atom? Let us digress here in order to learn how one expresses energies in atomic problems. We measure energy of atomic particles by a unit called "electron volt" (eV). It is the energy an electron would receive from a voltage (a change of potential) of one volt. The voltage is the "pressure" of electricity. For example, the outlets in our houses have a voltage pressure of 120 volts, which forces the current through our electric bulbs or appliances. If electrons could move freely between the terminals of an electric outlet of 120 volts they would leave the negative end of the terminal (electrons are negatively charged), fly across the gap, and arrive at the positive end with an energy of 120 electron volts. Actually they cannot leave the ends because they are kept within the confines of the metal by electric forces. Hence the current flows only if the ends are connected by a metal wire. If the ends are brought at close distance, they can overcome that restriction and bridge the gap. This is seen in the form of a spark, which develops if the two ends are brought close to one another.

The electron volt is a convenient unit of energy for our problems. For example, in air of ordinary temperature the molecules fly to and fro with an average kinetic energy of $1/40$ electron volts. This is the average energy per atom of any kind of heat motion at room temperature; it is, for example, the energy of the irregular heat oscillations that atoms perform in a piece of metal, the ones that cause melting at higher temperatures when the forces keeping the atoms in place are overcome.

Let us return to the experiments of Franck and Hertz, in which energy is transmitted to atoms by means of an electron beam. The threshold energy of a mercury atom—the minimum energy it is able to take in and add to its energy content—was found to be 4.3 electron volts; in the hydrogen atom it is as high as 10 electron volts. These are much higher energies than the energy of heat motion at room temperature. We immediately see a connection here with the fact that the atoms in a gas of room temperature maintain their identity and are not changed in spite of the many collisions they suffer. The energy of these collisions is below the threshold energy; that is, below the smallest energy quantum that the atom can accept. Thus the Franck-Hertz experiments showed in their own way the surprising stability of atoms and gave it a quantitative aspect. The atom remains unchanged and stable so long as the impacts upon it are less energetic than a certain well-defined threshold energy, and this energy has a characteristic value for each element. In effect, Franck and Hertz "measured" the atomic stability.

The results of experiments similar to those of Franck and Hertz go farther than this. They tell us not only the minimum amount of energy that the atoms would accept, they tell us the whole series of specific energy values the atom is willing to accept. Only these amounts of energy can be fed into the atom; it rejects anything in between. For example, the hydrogen atom accepts only the following amounts: 10 eV, 12 eV, 12.5 eV, and 12.9 eV, and higher values at decreasingly smaller intervals. The sodium atom accepts only 2.1 eV, 3.18 eV, 3.6 eV, 3.75 eV, etc. Figure 25 shows a graphic representation of these energies. Each energy corresponds to a

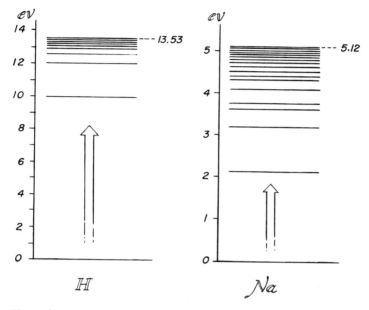

Figure 25
Energy of quantum states of hydrogen (H) and sodium (Na).

certain "mode of motion" of the electron in the atom. We call such a "mode of motion" a *state*. Hence each line represents a selected state that the atom is allowed to assume. All other states lying in between are seemingly forbidden. The selected states are called quantum states. The state of lowest energy is called the *ground state;* it is the state in which the atom is found normally. The other ones are called excited states. The difference between the energy of the first excited state and the ground state is the threshold energy.

These facts contrast sharply with what we expect from the behavior of the planetary model. Why should the energy of the electrons be quantized within the atom? What prevents us from adding an arbitrarily small amount to the energy of an atom? If one compares the energy of an atom with the size of a bank account, it is as if the bank would allow only certain prescribed amounts of money to be withdrawn or deposited, to keep the balance of the bank account at one of a series of predetermined values.

Let us now have a closer look at the different quantum states. One generally refers to the series of allowed energy values as the "term spectrum" of the atom. The two-term spectra in figure 25 reveal a very important general property of quantum states. The higher the energy lies above the ground state, the smaller the energy gap between quantum states becomes. It is a property observed in all atomic systems; for large excitation energies the quantum states become so close to one another that they almost merge. At high energies the quantum effects are not noticeable. The atom then is affected by any small amount of energy, as an ordinary planetary system would be. It is as if the strange rules regarding the bank account are waived for very large accounts so that the number of permitted deposits and withdrawals increases for bigger accounts.

This fact has turned out to be of fundamental importance and much more sweeping than it appears here. Today we know that when we pump enough energy into atoms, they behave as planetary systems do. These conditions are realized under extremely high temperatures, such as in strong electrical discharges in gases. Under these conditions the gas forms a so-called *plasma*.[21] In a plasma the atoms lose their characteristic properties; different kinds of atoms behave like one another.

21. The name *plasma* has nothing to do with blood plasma, or the living matter in a cell. The first realizations of an atomic plasma in a discharge tube looked like the biological plasma; hence this terminology.

For example, at room temperature the elements sodium and neon are about as different from one another as can be imagined. Sodium is a metal, one of the most chemically reactive elements; neon is a noble gas, called "noble" because it stands alone, forming no chemical compounds. Atoms of neon and sodium have different excitation energies: neon higher than sodium. But when very high energies are pumped into neon and sodium, both give off the same plasma radiation, which is a continuous spread of frequencies. Under these circumstances neither atom has discrete energies nor gives off light of unique frequencies. Individual atoms of sodium as well as neon can accept or emit any amount of energy, large or small; different kinds of atoms are no longer distinguishable from one another.

Chaos reigns in a plasma; it is a chaos of very high temperatures and is rarely found here on Earth except when produced in our laboratories. We find it, however, in cosmic space among the gases expelled by the sun and other hot stars.

In the plasma all orderly features by which we recognize one atom from another disappear. The continuous range of properties and the lack of distinction between different kinds of atoms in a plasma is just what one would expect on the basis of atoms as planetary systems, which would change continuously when colliding with one another. Order and differentiation occur only when the atoms are in their low-energy states, which are discrete and far apart on the energy scale. Then we find the stability that leads to specific shapes and orbits and, consequently, to specific chemical properties. It was these characteristic features at *low* energy that defied our understanding in classical terms and led to the more powerful picture of quantum mechanics.

## The Wave Nature of Atomic Particles

**Particle and Light Beams**    We now come to the most striking but most revealing group of observations. They deal with the nature of the atomic particles. Let us consider the simplest form in which atomic particles, say electrons, are found. This is when they are removed from atoms, and are freely moving in empty space. If the electrons in a stream are all moving in the same direction and with the same speed, we call the result an electron beam. Such beams are produced, for example, in television tubes. They hit the television screen from the inside of the tube and form the picture. Electron beams must be produced in a vacuum, since in ordinary air the electrons would bump into the air molecules and quickly get deflected out of the beam.

You might expect that such electron beams would have very simple properties, that they would represent a group of particles moving along parallel paths at the same speed and traveling in straight lines in free space. If they hit an obstacle, the particles would be scattered in all directions. On the contrary, however, we find very strange and unexpected phenomena.

Before we describe these effects, let us consider another kind of beam, a beam of light—the well-focused beam of a searchlight, for example. We assume that the light is of one color. Let us compare these two beams. We expect them to be fundamentally different things. The light beam is a bundle of electromagnetic waves propagating through space in a certain direction; no material is moving, only the state of the electromagnetic field in space is changing. In contrast, a beam of particles should consist of actual matter in small units moving straight forward. You would expect the two to be as different as the motion of waves on a lake from that of a school of fish swimming in the same direction.

Let us recall the experiments in which we have shown the wave nature of light, in particular the setup in which an obstacle is put in the way of the beam, as indicated in figure 14 of chapter three for light and in figure 26 for an electron beam. This setup seems to be ideal to bring out the difference between a beam of waves and a beam of particles. If the obstacle is put in the way of a beam of waves, we observe the characteristic interference patterns shown in figure 14. If the obstacle is put in the way of a particle beam, we would expect the particles that hit it to miss the screen; the ones missing the obstacle would reach the screen; the ones just passing by at the edge of the obstacle might be scattered and deviated from their path. Hence, if we use a screen of the same material of which television screens are made, we should observe a region of shadow and a region of light, the transition being not quite sharp because of the scattering at the edge. No stripes are expected when no wave phenomenon is involved.

What a surprise for the physicists when they performed this and similar experiments and found electron beams exhibiting wave properties similar to those of light beams! Figure 27 shows the pattern an electron beam formed on a screen in the arrangement of figure 14. The pattern is identical with the one of figure 16 observed with light. This amazing result is only one of many that have shown beyond doubt that electron beams must have some kind of wave nature. There must be a wave involved in the electron motion.

A quantitative study of these interference patterns allows one to measure the wavelength of this mysterious "electron wave." The wavelength depends on the speed of the electron—the higher the speed, the smaller the wavelength; for electrons with an energy of a few electron volts, the wavelength is of the size of the atoms. It is a very small wavelength indeed, and this is why the wave nature of elec-

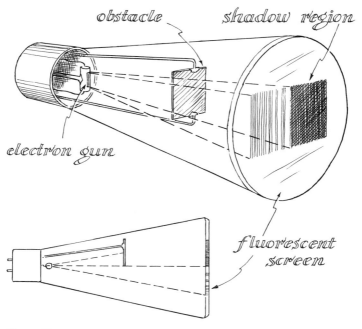

Figure 26
Electron beam diffraction apparatus, analogous to light diffraction apparatus in figure 14. For the sake of clarity, the pattern of stripes is drawn much wider than it would be with actual electrons.

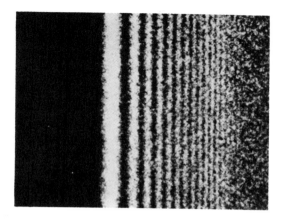

Figure 27
Interference of electrons is demonstrated by this actual photograph, which is analogous to the phenomenon of figure 16. (From H. Raether, "Electroninterferenzen," *Handbuch der Physik* 32 (Berlin: Springer, 1957).)

tron beams is not easy to detect. In most practical applications of electron beams, such as television tubes, the wave nature plays no role whatsoever.

Many experiments had to be performed before the physicists were really convinced that the wave effects were not caused by some other phenomenon. All these experiments, however, only made it more and more clear that waves play a part in the motion of electrons and also of other atomic particles, such as nuclei.

An obvious question poses itself: How can an electron be a particle and a wave at the same time? A wave is something that is spread over space in a continuous way, but a particle is strictly localized; at a given moment the particle is here and not there, whereas a wave is a state of "tension" of space that must spread over at least a few wavelengths in order to represent something that can be called a wave. Can we perform some decisive experiment to settle the question unambiguously? Is the electron really a particle or a wave?

Of all the questions that have been raised by twentieth century physics, this is perhaps the most interesting. But before we discuss it, we must be aware of the most exciting thing regarding the electron waves: The dual nature of electrons as particles and waves contains the clue to the riddle of atomic structure! The unexpected properties of the electrons circling around atomic nuclei are directly connected with their wave nature.

**The Properties of Confined Waves**  In order to understand the connections between electron waves and atomic properties, we must first study the peculiar behavior of waves when they are confined to a limited region.

Let us take the simplest example, of waves along an extended rope. If the rope is very long, we can produce a wave running along the rope by shaking one end, as every child playing with a skip rope knows well. If the rope is tied at its far end to a fixed object and held under tension, the impulse travels along the rope and sometimes returns to us after being reflected at the point where the rope is attached. With practice, we can shake one end of the rope to produce any form of wave, with long or short wavelength, just as we wish. The long wavelengths yield slow oscillations; in the short ones the rope will vibrate quickly when the waves pass by. Now let us confine the rope between two nearby points. It is then better to think not of a rope but of a string that is suspended under tension between two points, such as a string on a violin. The form of vibration of such a string is what is called a *standing wave*. We no longer have the choice of wavelength and frequency; in fact only those vibrations can be set up whose half wavelength fits once or twice or any integral number of times into the space between the two points of attachment, as shown in figure 28.

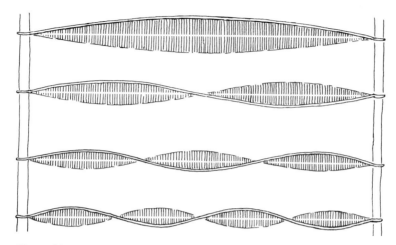

Figure 28
Standing waves. Vibrations of a string confined between the points of attachment. Only vibrations set up in which one, two, three, four, etc., half wavelengths fit in the space between the attachments. Straight white line indicates position of the string at rest.

Not only are the shapes of the vibrations determined, but also the frequencies (the number of ups and downs per second), once the tension of the string is kept fixed. Each of the different vibrations that can be set up has its characteristic frequency, so the string can vibrate only with a set of given frequencies. The lowest of these frequencies, the easiest to set up, is the one whose half wavelength just fits the distance between the fixed ends of the string. It is the one the violinist plays when he sets the string in motion with his bow. But he can also set up higher vibrations, the so-called flageolet tones, in which two or more half wavelengths fit in the string. Even when he plays a normal tone, the motion of the string is not purely the lowest vibration. The actual motion is a combination of several permitted forms of motion. In fact the ordinary musical tone of a violin contains the higher modes to a certain degree; they are the harmonics, whose presence is important for the beauty of the sound. The difference between the tone of Haifetz and that of an ordinary violin player results from the different admixture of higher modes. But whatever the combination is, it can contain only frequencies from the set assigned to the tuned string.

The lesson learned from the string generally is true for all kinds of waves. Whenever waves are confined to a finite space, we observe special wave forms and a set of assigned frequencies that are characteristic of the system. Most musical instruments are built on this principle. The string instruments make use of the series of discrete and characteristic frequencies of the string; a wind instrument is based upon the assigned frequencies of air waves enclosed in the pipe of the instrument, be it a trumpet or an organ pipe.

Another interesting example of this phenomenon can be seen when water waves are confined, say, in a glass of water. You can observe a very striking example by rubbing your finger around the rim of a stemware glass (such as a wine or brandy glass) that is partially full of water. If your finger is free of oil, it will stick and slip around the rim like a bow over a violin string, causing a vibration that sets up a pattern of surface waves in the glass. This pattern is accompanied by a flutelike tone that has been used in the novelty instrument called the "glass harmonica." If the amount of water in the glass is changed, a different pattern will result on the surface of the water, accompanied by a different tone. You will be hearing characteristic frequencies connected with characteristic wave patterns.

It would be entirely possible to calculate the shape of these patterns and predict at what frequency of vibration they are supposed to appear. All that one must know is the size and shape of the glass and the properties of waves on the water surface.

**Electron Waves and Quantum States**    Let us now return to the electron waves. How can one confine electron waves and observe similar phenomena? Any situation that confines electrons will also confine electron waves. Such a situation exists when an electron is close to an atomic nucleus. The positive charge of the nucleus attracts the electron and prevents it from leaving the immediate neighborhood of the nucleus; the electron is confined to a space near the nucleus. What effect will this confinement have on the electron wave? This question was asked and answered first by Erwin Schroedinger in 1926.

He was able to calculate the shape and the frequencies of the characteristic patterns that develop when electron waves are confined by a nucleus. It is a straightforward problem of dynamics of confined waves, once the relation between the wavelength of the electron wave and the velocity of the electron is known. The result is a series of distinct vibrations, each of them with a characteristic pattern and frequency. The wave nature of the electron immediately "explains" the fact that the electron can assume only certain well-defined modes of motion in the atom, the now-familiar quantum states.

This result is of fundamental significance. A connection was found between the wave nature of the electron and the existence of discrete states in the atom. Here we are touching the very nerve of nature. When an electron is confined to a limited region around the nucleus, the wave properties of the electron permit only certain special, predetermined states of motion. Therefore the atom cannot change its state continuously; it must change abruptly from one allowed state to the other.

The success of the electron wave picture of the atom is all the more remarkable because of the way it fits all facts in every quantitative detail. Schroedinger first calculated the simplest problem, the hydrogen atom, in which one single electron is confined by the nucleus. He found a series of states of vibration that correspond in every respect to the observed quantum states of the hydrogen atom. For example, the extension in space of the waves corresponded very well with the observed size of the hydrogen atom. But the most surprising and convincing result of his calculations dealt with the energy of the quantum states. How do we get at the energies of the different states of vibration? Here we must introduce one of the most fundamental relations revealed by the new quantum mechanics, the relation between frequency $\nu$ and energy. Since the beginning of our century physicists have noticed a relation that is the basis of the wave-particle duality; wave frequencies and particle energies are intimately connected by a simple formula. The energy $E$ is the frequency $\nu$ multiplied by a famous number, Planck's constant, always referred to by the letter $h$. The formula reads

$$E = h\nu$$

and is called Planck's formula. Planck's constant $h$ is a small number; if one measures energies in electron volts and frequencies in ups and downs per second, the numerical value of $h$ is $4 \times 10^{-15}$. A vibration of $10^{15}$ times per second corresponds to 4 electron volts.

Now comes the great surprise. When Schroedinger calculated the states of vibration of an electron wave confined by the electric attraction of the nucleus, the frequencies of these vibrations corresponded exactly to the observed energies of the quantum states of hydrogen when the formula of Planck is used to convert frequencies into energies! This success is incredible: Schroedinger calculated the vibrations of an electron wave confined by the attraction of the center. He multiplied the frequencies by Planck's constant and obtained, exactly to the last decimal point, the energies of the quantum states of hydrogen, the allowed values of the energy bank account of the hydrogen atom.[22] Obviously the wave nature of the electron must be a decisive factor in the understanding of atomic properties.

The confinement of electron waves admits a series of possible states and furnishes a set of assigned frequencies. If we keep in mind the fundamental law connecting frequency with energy, we obtain a series of states with assigned energies. The one with the lowest frequency is the most important one, since it is the quantum state of lowest energy, the ground state of the atom. It is also the one that exhibits the wave nature most prominently.

We now understand what was observed in the Franck-Hertz experiment. The atom will stay in the state of lowest energy until it gets enough energy to be lifted into the next state. If the energy is insufficient, the atom cannot accept *any* energy and remains in the ground state.

The confined electron waves in atoms cannot be observed directly. We can measure their extension, their frequencies (to be exact, the differences between frequencies, which are observed as energy differences), and other indirect properties. But it is both constructive and impressive to look at pictures of these wave patterns. The pictures are not photographs; this would be impossible, as we shall see later in more detail. They are models made from the results of calculations. Figure 29 shows the electron wave patterns, in the order of increasing frequency or energy, of successive quantum states of an electron confined by a nucleus. The lowest state, the

22. Everyone who contemplates this fantastic discovery sympathizes with the famous Italian physicist Enrico Fermi, who used to say when presenting this calculation in his lectures with his well-known Italian accent, "It has no business to fit so well!"

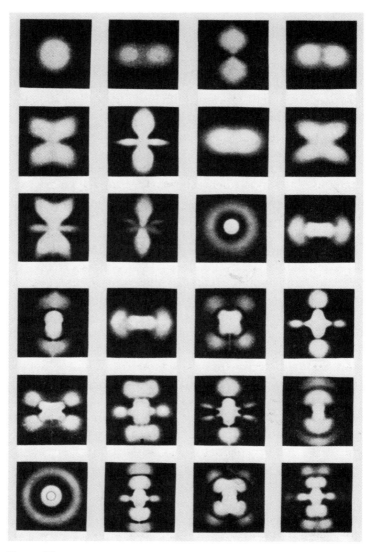

Figure 29
These are photographs not of real electrons but of models carefully constructed according to our observations and calculations. (From H. E. White, *Physical Review* 37 (1931).)

ground state, is the simplest one; the higher the frequency, the more involved the pattern. The ground state has spherical symmetry. The next ones have a figure-eight form. The higher ones are usually more complex, although we also find simpler ones among them.

These patterns are of utmost importance in the makeup of nature. They are the fundamental forms on which matter is built. They are the shapes, and the *only* shapes, that the electron "motion" can assume under the conditions prevailing in atoms—that is, under the influence of a central force (the attraction of the nucleus) that keeps the electron confined. Hence these patterns are the symbols of the way in which nature combines and forms everything we see around us.

The patterns of figure 29 and their inherent symmetries determine the behavior of the atoms; they are the basis of the orderly arrangement in molecules and also of the symmetric arrangement of atoms or molecules in crystals. The simple beauty of a crystal reflects on a larger scale the fundamental shapes of the atomic patterns. Ultimately, all the regularities of form and structure that we see in nature, ranging from the hexagonal shape of a snowflake to the intricate symmetries of living forms in flowers and animals, are based upon the symmetries of these atomic patterns.

Looking at the patterns, we see that the higher we go in frequency (or energy), the finer the pattern becomes, the smaller the distances between the ups and downs. The wavelength becomes shorter. If one goes to very high frequencies (energies), the pattern is so varied and fine-grained that it looks almost smooth and continuous. Hence the motion it describes will be very close to that of an ordinary particle without wave properties. Here again we see that our wave picture reproduces what we have found in atoms. When the energy is high, the quantum phenomena cease to be important and the atom behaves as if it were an ordinary planetary system. The transition to the plasma conditions at high energy is also contained in the wave nature of the electron.

The hydrogen atom in its ground state vibrates in the simplest possible pattern, the first one in figure 29. Other atoms, however, exhibit the more complex patterns even in their ground states. This is explained by an important principle, first discovered by Wolfgang Pauli in 1925. When more than one electron is confined in an atom, each electron must assume a different pattern. Thus an added electron will have to assume the next higher pattern in the scale.

Actually, up to two electrons are allowed in one pattern. The electrons have an interesting property: They rotate around their own axis. This rotation is called the electron spin. There are two possible states of rotation, either clockwise or counterclockwise. Thus any electron pattern counts twice, once with the electron rotating one way, and then with the electron rotating the other way. As we will see later in the chapter on chemistry, a singly occupied pattern has very different chemical properties than a doubly occupied one.

Here we find the explanation of the fact that one electron added or removed makes so much difference in the atomic world. The last electron added goes into an outer wave pattern, and this outer pattern is what neighboring atoms "see." This, in turn, determines the way the atoms fit together, whether they form a crystal, a liquid, or a gas. Therefore, the properties of the atoms change appreciably when going from one number of electrons to the next higher one. Quantity becomes quality in the atomic world; one electron more may lead to a complete change of properties.

Schroedinger's discovery of the fundamental significance

of the electron wave for the structure of the atom, and the development of the theory by Heisenberg, Max Born, and Pauli, marks a turning point in man's understanding of nature comparable to Newton's discovery of the universal nature of gravity, Maxwell's electromagnetic theory of light, and Einstein's relativity theory. The properties of the atoms, which seemed so strange and incomprehensible on the basis of the planetary model, fall into place when considered in the light of a confined wave phenomenon. A confined wave assumes certain well-defined shapes and frequencies, such as the vibration of the air in an organ pipe, the string on a violin, or the water surface in a vibrating glass. They all form a series of vibrating patterns, beginning with the simplest pattern, which vibrates with the lowest frequency, and including more complicated patterns of higher frequencies. Vibrations of electron waves in atoms have the same properties.

With this new way of looking at nature, we now can understand the three remarkable properties of the atom we enumerated at the end of the last chapter. The *stability* comes from the fact that considerable energy must be added to change the lowest pattern to the next higher one.[23] As long as the influences on the atom are less energetic than this energy, the atom remains in its lowest pattern. The configuration of the atom, therefore, exhibits the typical stability. The *identity* of atoms comes from the fact that the wave patterns are always the same and are determined by the way the waves are confined. One sodium atom in its lowest energy state is identical to another because the electron wave is confined by the same conditions—that is, by the attraction of the nucleus and the electric effects of the other electrons in the atom. Two

23. According to Planck's formula, this energy is equal to the frequency difference multiplied by Planck's constant.

gold atoms are identical because the same number of electrons are confined by the same electric charge in the center
and, therefore, produce the same wave vibrations. Finally, the
ability of an atom to *regenerate* its original shape after distortion is what one expects of a vibrational wave phenomenon,
for the same reason that two of the same kind of atoms are
identical. When the original conditions are reestablished, the
electron vibration again assumes the same pattern as before,
since the patterns are uniquely determined by the conditions
in which the electron moves and are quite independent of
what happened before. The patterns do not depend at all on
the previous history of the atom; we may destroy an atom by
removing a few electrons or distort it by condensing the material into a solid, as we did in the example of sodium in the last
chapter, but whenever we get the atom back into the original
conditions, the electron waves will assume the same quantum
states they had before. There exists only one wave pattern of
lowest frequency or energy for each atom.

It is remarkable that we actually find in the world of atoms
what Pythagoras and Kepler sought vainly to find in the motion of the planets. They believed that the earth and other
planets move in special orbits, one unique to each planet and
determined by some ultimate principle that is independent of
the particular fate and past history of our planetary system.
We have found no such principle in the motion of planets, but
we do find one in the motion of electrons in atoms—the wave
principle. We are reminded of the Pythagorean harmony of
the world: The atomic quantum states have specific shapes
and frequencies that are uniquely predetermined. Every hydrogen atom in the world strikes the same chord of vibrations, as given by its set of characteristic frequencies. Here we
find the "harmony of the spheres" reappearing in the atomic
world, but this time clearly understood as a typical property
of confined electron waves. (See figure 30.)

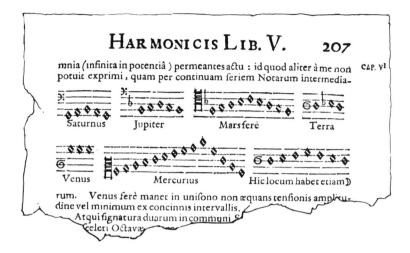

Figure 30
Harmony of Spheres (from *Harmonicus Mundi* by Johannes Kepler, 1619).

## The Light Quanta

**The Graininess of Light**  We have learned that electrons and other atomic particles exhibit wave properties. Particle beams sometimes behave as if they were waves. This property was shown to be the basis of the quantum behavior of the atoms. It turned out in the course of research that this duality is not restricted to particles only. Incredible as it may seem, the light we described as waves in chapter 3 is found experimentally to behave sometimes as if it were particles.

Many observations regarding the propagation of light indicate that a light beam is a continuous wave of electromagnetic fields. But when the effects of light on matter were studied more closely and with more sophisticated tools, some unexpected phenomena were observed that seemingly contradicted the picture of a continuous flow of light. What happens if light falls on matter? If the object is transparent, such as a windowpane, light is partially reflected and partially transmitted. If the object is opaque, such as a piece of coal, or partially transparent, such as colored glass, a good part of the light is neither reflected nor transmitted; it disappears into the object. Since light is a form of energy, it can disappear only by giving its energy to matter in some way. This disappearance is called the absorption of light.

The energy of the absorbed light must show up in some other form. We feel the heat when sunlight is absorbed by our skin. When light is absorbed by some metals, its energy is often transferred to electrons, which then have acquired so much energy that they jump out of the metal. This jumping is called the photoelectric effect, which is of practical use when we want to transform light pulses into electrical pulses.

It is possible to measure with great accuracy the energy transferred to matter when light is absorbed. These measurements have had a most unexpected result. Light energy can be absorbed only in definite units of a certain amount; a fraction of these units can never be absorbed. If we compare energy with money, we might say that a light beam transmits its energy to matter only in full dollars but never in small change. The units are called light quanta, or photons. As far as the effect of light on matter is concerned, we can compare a light beam to a stream of bullets, each filled with the same amount of explosive. Whenever a bullet hits an object, it causes an effect whose energy is determined by the amount of explosive. Stronger light means more explosions of the same size, but not stronger explosions.

In the photoelectric effect each light quantum hitting the metal forces an electron to jump out of the metal. The energy of the jumping electron is a measure of the size of the light quantum (it measures the amount of explosive in each bullet). The electrons that are forced out of the metal by light are called *photoelectrons*. The *number* of electrons jumping out determines the intensity of the light beam, which depends on the number of quanta entering the metal. However, the *energy* of the individual photoelectrons emitted depends only on the amount of energy in each incident light quantum—the amount of explosive in the bullet.

The amount of energy in the light quantum depends on the color or frequency of light we are dealing with. It is different for light of different frequencies: higher frequencies have larger units of energy, lower frequencies have smaller ones (figure 31). The energy quantum of visible light is small. It contains an energy of only a few electron volts, about $10^{-12}$ (a millionth of a millionth) smaller than the energy of a touch on your finger that you can barely feel. If this energy quantum were not that small, the quantum behavior of light would have been discovered much earlier. Of course, the retinas in our eyes are much more sensitive to visible light than our finger tips. Still we are unable to see single light quanta be-

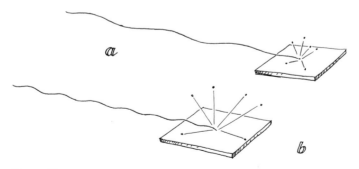

Figure 31
Photoelectric effect. (a) Low frequency light (long wavelength) impinging on metal ejects slow electrons. (b) High frequency light (short wavelength) ejects fast electrons.

cause they are too weak. If we could see them, a very weak light source would appear as an intermittent light, since we would see light only when a quantum arrived at the retina.

Although light is an electromagnetic wave, its effect on matter, on our eyes, on the photoelectric cell, is quantized; the light beam consists of small grains, each of the same size. This phenomenon emphasizes the particle-wave duality in nature: Electrons are particles with wave properties; light is a wave with particle properties.

Let us be a little more quantitative. The size of the energy quantum of light is connected to the frequency of light by the same formula we already have encountered, the formula of Planck. The energy $E$ of a quantum is given by $E = h\nu$, where $\nu$ is the frequency of light,[24] and $h$ is again Planck's constant.

Small as these quanta are, they are not small amounts of energy compared to the energies of atoms. They are of the same order of magnitude as the energy of the atomic quantum states. For example, a quantum of yellow light is about two electron volts. This is close to the energy (2.1 eV) necessary to lift the sodium atom from its ground state to its next higher state.

**Atoms and Light Quanta**    Strange as the idea of the light quantum is, it opens up a new aspect of how an atom emits and absorbs light, how light is produced by atoms, and how atoms are influenced by light. Let us combine the concept of the light quantum with the concept of the quantum states of the atom. We have learned that an atom can be found only in certain quantum states, with definite energies characteristic of

24.  The frequency of a light wave is the number of vibrations of the electric field per second. Long wavelengths correspond to low frequencies; short wavelengths have high frequencies. Ordinary radio waves have frequencies of about $10^6$ per second; visible light has frequencies of about $10^{15}$ per second.

the type of atom. Thus an atom can gain or lose energy only in amounts corresponding to the energy differences between its quantum states. If the atom absorbs light or emits light, the energy of this light must be equal to one of those differences. Hence the atom can emit or absorb only light with quanta of the correct amount—namely, an amount equal to one of these differences.

This property explains immediately why atoms radiate and absorb only light with certain typical frequencies. For example, an atom in its ground state can accept only light whose quantum energy is just the right size to lift the atom into one of the higher quantum states. The same is true of light emission. Light can be emitted by an atom only when the atom is in a state higher than the ground state, and then it can emit only light whose quanta correspond to the energy difference between that state and a lower state. The atom can give off or take in only light quanta such that the energy balance will leave it again in a quantum state. Hence any light absorbed or emitted by an atom must have a frequency corresponding to the difference of two characteristic energy values.

Let us take the sodium atom as an example. When sodium gas is cool, all atoms are in the ground state. No radiation is emitted. The gas is transparent to light, except for light whose quanta would lift it to a higher state. (Such higher states we have called *excited states*.) For example, according to figure 25, the first excited state is 2.1 eV higher than the ground state; 2.1 eV corresponds to a frequency of $5.2 \times 10^{14}$ vibrations per second. Hence light of that frequency—yellow light—will be absorbed by sodium gas.

Instead of absorption we can have emission. All we have to do is pour energy into the sodium gas, either by heat or by an electric discharge, as is done in the yellow sodium street lights along some of our highways. Then a few sodium atoms will be put into higher quantum states. Those atoms then emit light when they fall back into a lower state. The ones in the first excited state emit the same yellow light that the cool gas has absorbed. It is the color we see radiated by these street lights. When the temperature or the discharge energy is raised, higher and higher quantum states will be created and more colors will appear.

It is most remarkable how well the results of light-radiation experiments fitted the results of the Franck-Hertz and similar experiments. In light-radiation experiments you measure frequencies; in experiments of the Franck-Hertz type you measure energy differences between quantum states. Without exception, all frequencies emitted and absorbed by atoms correspond to energies of transition from one quantum state to another.

## The Complementarity between the Particle and Wave Pictures

Now we must return to our fundamental question: How can an electron be a particle and a wave at the same time? It is difficult to formulate the answer to this question in simple terms. The unexpected dual characteristic of matter has shown that our ordinary concepts of particle motion are not adequate for a description of what goes on in the atomic world. After all, these concepts are formed from human experience with visible objects that are larger than the atomic particles by factors of many millions. In order to understand what is going on at the atomic scale, we must be prepared to give up accustomed ways of thinking and replace them with new concepts that nature has forced on us.

One of the features of classical physics that we must question is the "divisibility" of atomic phenomena. This is the idea that a physical process can be thought of as consisting of a succession of particular processes. According to this idea, theoretically at least, each process can be followed step by step in time and space. The orbit of an electron around the nucleus would be thought of as a succession of small displacements. Is this kind of description consistent with what one finds within the atoms?

In our ordinary way of looking at things, the electron must be either a particle or a wave; it cannot be both at the same time. After all, a careful tracing of the electron along its path must decide this question and put it in either one or the other category. Here the problem of the divisibility of atomic phenomena comes in. Can we really perform this tracing? There are technical problems in the way. If we want to "look" at the detailed structure of the orbit, we must use light waves with very small wavelength. Such light, however, has a high frequency, hence a large energy quantum. In fact light whose wavelength is as small as an atomic orbit has a quantum of energy that would be far more than enough to tear the electron away from the atom. When it hits the electron, it will knock it out of its orbit and destroy the very object of our examination.

Destruction of the object of examination is not peculiar to experiments in which light is used to trace the electron orbit. Quite generally, all measurements that could be of use to decide between the wave and the particle nature of the electron (or the proton, or any other atomic entity) have the same outcome. If one performs these measurements, the object changes its state completely. Indeed the state into which the object is put by the measurement is a very high energy state, which therefore does not show any wave properties.

The quantum nature of light or of any other means of observation introduces a coarseness that makes it impossible to decide between wave and particle. It does not allow us to subdivide the atomic orbit into a succession of partial motions, be it particle displacements or wave oscillations. If we force a subdivision of the process and try to look more accurately at the wave in order to find out where the electron "really" is, we will find it there as a real particle, but we will have destroyed the subtle individuality of the quantum state. The wave nature will have disappeared, and with it all the characteristic properties of the atom. After all, it was the wave nature that gave rise to the typical properties of quantum states—the simple shape, the regeneration of the original form after perturbation, and all other specific qualities of the atom.

The wave nature of the electron is predicated on the indivisibility of the quantum state. The great new insight of quantum physics is the recognition that the individual quantum states form an indivisible whole, which exists only as long as it is not attacked by penetrating means of observation. In the quantum state the electron is neither a particle nor a wave in the old senses. The quantum state is the form an atom assumes when it is left alone to adjust itself to the conditions at low energies. It forms a definite individual entity, whose pattern and shape correspond to a wave motion, spreading out over a finite region of space. Any attempt to look at its detailed structure by direct observation would unavoidably destroy it, since the tools of observation would pour so much energy into the system that the condition of low energy would no longer hold.

At this stage of our discussion it will appear quite natural that predictions of atomic phenomena sometimes must remain probability statements only. The prediction of the exact spot where the electron will be found when the quantum state is destroyed with high energy light is a case of this kind. If the quantum state is examined with pinpointing light, the electron will be found somewhere in the region of the wave, but the exact point cannot be predicted with accuracy. Only probability statements can be made—for example, that the electron will be found most probably where the electron wave was most intense.

The central place for probability statements in describing atomic particles is the famous uncertainty principle of Heisenberg. Let us look at its formulation concerning velocity and position measurements. Imagine that we have before us a thousand or a million atoms, all equal and all in the same quantum state—for example, in the ground state. Let us now measure the position or velocity of an electron in each of these atoms. We will not always get the same result, although the atoms are supposed to be in identical quantum states. There will be a spread in the velocity measurement and in the position measurements. The uncertainty principle says that, whatever the nature of the atoms, the product of the spreads must be larger than a certain number, namely $h/m$; $h$ is the famous constant of Planck, which we have encountered before, and $m$ is the mass of the electron. Indeed this principle holds even if the measurements are made not only on atoms in the same quantum state but on electrons or any other particles in any possible systems or configurations (free electrons, electrons confined in a box, etc.), as long as they are made on a large number of equal systems or configurations. This numerical value of the product of the spreads is just big enough to make it impossible to decide between the wave and the particle picture of the electron. Clearly, if both position and velocity had exact and definite values in a given quantum state, the electron could not be a wave, since a wave of a given wavelength (the wavelength is determined by the velocity) is necessarily spread out and cannot have a definite position (here and nowhere else). Each individual position measurement and each velocity measurement may be made as accurately as you wish; however, the uncertainty principle does not describe individual measurements, but rather the probabilistic *spread* of these measurements when made on a large number of systems in which electrons are all in the same quantum state. And the relative spreads are just large enough so that the wave nature of the electrons (or any other particles) cannot be disproved.

This principle expresses a negative statement. We must recognize, however, that the highly important fact of the spread in either position or velocity measurements in a given quantum state is more than a mere technical limitation that some day might be overcome by clever instrumentation. If it were so, the coexistence of wave and particle properties in a single object would collapse, since it would prove the wave picture to be wrong. We know from a great wealth of observations that our objects exhibit both wave and particle properties. Hence the Heisenberg restrictions must have a deeper root: They are a necessary corollary to the dual nature of atomic objects. If they were broken, our interpretation of the wide field of atomic phenomena would be nothing but a web of errors, and its amazing success would be based upon accidental coincidence.

Quantum mechanics has given us an unexpected but wonderful answer to a great dilemma. On the one hand, atoms are the smallest parts of matter; they are supposed to be indivisible and endowed with every detailed specific property of the´substance. On the other hand, atoms are known to have an internal structure; they consist of electrons and nuclei, which, if prequantum physics were applicable, would perform mechanical motions not unlike the planets around the sun, and therefore cannot be imagined to exhibit the required properties.

The answer lies in the discovery of the quantum states,

which solves, to some extent, the first part of the dilemma. Wavelike behavior endows atoms with the properties of identity, wholeness, and specificity, but the range of this behavior is limited. They will retain their identity and their specific properties as long as the perturbances to which they are exposed are smaller than a characteristic threshold. If they are exposed to stronger perturbations, the atoms lose their typical quantum properties and exhibit the untypical behavior expected from the mechanical properties of their internal structure.

The quantum state cannot be described in terms of a mechanical model. It is a phenomenon different from what we have experienced with large objects. It has a particular way of escaping ordinary observation because of the fact that such observation necessarily will obliterate the conditions of its existence. The great Danish physicist Niels Bohr, who has contributed most to the clarification of these ideas, uses a special term for this remarkable situation; he calls it *complementarity*. The two descriptions of the atom—the wavelike quantum state on the one hand, and the planetary model on the other—are complementary descriptions, each equally true but applicable in different situations.

The quantum properties can unfold only when the atom is left undisturbed, or when it is exposed to perturbations that are less energetic than the quantum threshold. Then we find the atom with its characteristic symmetries, and it behaves like an indivisible entity. This is the case when we are dealing with matter under normal conditions. But when we try to look into the details of the quantum state by some sharp instrument of observation, we necessarily pour much energy into the atoms. Under these conditions the atoms behave as they would at very high temperature; that is, as a plasma. We then observe the electrons as ordinary particles moving under the attractive force of the nuclei, without any quantum phenomena, and exactly as one would expect if one had to deal with ordinary, old-fashioned particles.

Atomic phenomena present us with a much richer reality than we are accustomed to meeting in macroscopic physics. The wavelike properties of quantum states, the indivisibility of these states, the fact that we cannot describe the atom completely in terms of familiar things such as particles or classical waves, are features that do not occur with objects in our macroscopic experience. Hence the description of the atom cannot be as "detached" from the observing process as prequantum descriptions were. We can describe atomic reality only by telling truthfully what happens when we observe a phenomenon in different ways, although it may seem incredible to the uninitiated that the same electron can behave so differently as we observe it in the two complementary situations. These features, however, do not make electrons less real than anything else we observe in nature. Indeed the quantum states of the electron are the very basis of what we call reality around us.

# 6 Chemistry

## The Chemical Bond

In the preceding chapter we examined the structure of the atom and saw how the wave pattern of the electrons endows each atom with its typical properties. We looked at each atom as a single unit by itself, but we did not consider what happens when several atoms come close to each other. We do know, however, that the smallest units of many materials are not atoms but molecules, which are groups of atoms closely bound to each other. If we are to understand the structure of matter, we must understand not only the structure of atoms but also the reasons why atoms join and form molecules. We must understand what is called the *chemical bond,* which keeps the atoms together within the molecule, and we must become acquainted with a few typical molecules and their properties. The chemical bond and the properties of molecules are the subjects of chemistry.

Before the advent of quantum mechanics it was believed that there existed a special "chemical force" responsible for the chemical bond. It would have to be a most peculiar force, since some groups of atoms stick together very well, others not at all. For example, once two hydrogen atoms have assembled into a unit, the hydrogen molecule $H_2$, no additional atoms can be added to the group. The molecule is saturated. There is no molecule $H_3$ consisting of three hydrogen atoms. The chemical forces seem to have disappeared in respect to additional atoms.

Quantum mechanics has given a complete explanation of the chemical phenomena. There is no new force in action here at all. The chemical bond between atoms is the effect of the interplay of the electronic patterns of different atoms. A chemical bond arises when the patterns fit well together, like the cogs of the driving shafts in a gear or the pieces of a jigsaw puzzle. The patterns enmesh and interlace when the atoms are brought into contact; they merge into new wave patterns.

Some atomic patterns fit well together, others not so well. The chemical bond depends very much on the kinds of atoms involved. Sometimes they fit so well that when some atoms, such as those in figure 32a, are brought together, the patterns merge into one unit, only somewhat deformed and tighter, as shown in figure 32b. The assembly then forms a saturated molecule, which will not bind any further atoms. A saturated molecule can be compared to a finished jigsaw puzzle, where all parts merge into one unit and there is no place for an additional part.

Since it comes from the combination and interlacing of electronic wave patterns, the chemical bond is fundamentally electric in its nature. Its strength is due to the quantum stability of the combined electronic wave patterns in the molecule that are held together by the electric attraction of both nuclei. The various patterns in figure 29 illustrate the many ways in which these patterns can be combined and enmeshed. Consequently, we expect many different types of chemical compounds.

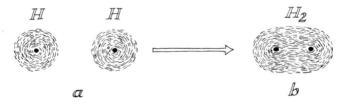

Figure 32
Two hydrogen atoms, each having the simplest electron pattern of figure 29, form a hydrogen molecule, where the two electrons merge into one elliptical pattern. The dots in the center are the hydrogen nuclei.

## Molecules

Let us look at a few specific examples. Among the many ways atoms hold together, there are three important kinds of bonds. One is the *twin-electron* bond, the second is the *plug-and-hole* bond, and the third is the *ionic* bond. The characteristic example of the first bond is the simplest molecule of all, the hydrogen molecule $H_2$, which consists of only two hydrogen atoms. Here the two electrons, one from each atom, merge into one pattern. This merging keeps the atoms together (figure 32); however, it seems to contradict Pauli's principle, which says that not more than one electron can assume a given pattern. But as we have already learned, two electrons can share the same pattern, provided that their spins are opposed to each other. This is why the electron spin plays such an important role in conjunction with the Pauli principle—it makes it possible for two, but only two, electrons to merge into one pattern. We can make a molecule out of two hydrogen atoms by having their electrons merge in a common pattern, but we cannot make a molecule out of three. The chemical binding is saturated with two electrons in one orbit. This is what we call a *twin-electron* bond.

Another interesting example of twin-electron bonds is the ammonia molecule $NH_3$, formed of a nitrogen atom and three hydrogen atoms. The nitrogen atom has seven electrons, of which four form a tight spherical pattern around the nucleus; the remaining three form a pattern of prongs extended in three directions perpendicular to each other, forward, to the side, and upward. (See figure 33a.) With this picture, we can easily understand the structure of the important molecule of ammonia $NH_3$, in which three hydrogen atoms form twin-electron bonds with each of the prongs. The electrons of the hydrogen atoms merge with the electrons in

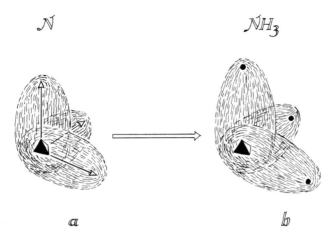

Figure 33
(a) The nitrogen atom. There are three electron prongs in the directions of the arrows at right angles to each other. (b) The ammonia molecule. Each electron prong of nitrogen merges with a hydrogen electron and forms a twin-electron pair. The heavy triangle is the nitrogen nucleus. The small black circles are hydrogen nuclei.

the prongs and we get a structure of the form indicated in figure 33b, where the hydrogen nuclei sit at the tips of the nitrogen prongs.

The carbon atom is particularly suited for molecule formation. It has six electrons; two of them are close to the nucleus in the form of a small round pattern, and four can arrange themselves in a symmetrical pattern in which each forms a radial prong away from the center. The tips of the four prongs form a regular tetrahedron (figure 34a), and this picture enables us to understand the arrangement of the molecule of methane, $CH_4$, the main component of cooking gas. It consists of one carbon atom and four hydrogens. The electrons of the hydrogen atoms merge with the four prong pat-

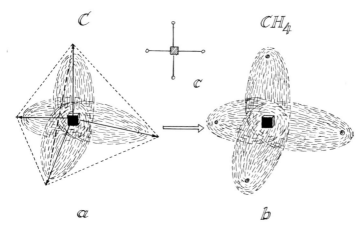

Figure 34
(a) The carbon atom with its four electron prongs directed at the four corners of a regular tetrahedron. (b) The methane molecule $CH_4$. Each electron prong of the carbon atom merges with a hydrogen electron and forms a twin-electron bond. The dark square is a carbon nucleus. The small circles are hydrogen nuclei. (c) Schematic figure of $CH_4$. The twin-electron bonds are indicated by lines.

terns in twin-electron bonds, creating a structure in which the carbon nucleus is in the center and four protons are at the corners of a tetrahedron (figure 34b).

Another kind of chemical bond, another way in which wave patterns merge, is the *plug-and-hole* bond. A characteristic example of this bond is the water molecule $H_2O$, which consists of two hydrogen atoms and one oxygen. The oxygen atom has eight electrons. Now it happens that ten electrons around an atomic nucleus form an assembly in a very tight, round pattern. The element neon, which has ten electrons, is a good example of this. Its ten electrons form such a closed tight pattern that it is chemically very inactive and does not form any molecules. In the oxygen atom with eight electrons, two electrons are missing from the tight pattern. Thus one can describe the pattern assembly of eight electrons as a tight round pattern with two "holes" in it. The holes have a well-defined shape, the shape of a missing electron pattern. In the case of oxygen the holes extend from the surface to the center, and the two holes are directed at right angles to each other. (See figure 35a.) Now we can understand the structure of the water molecule. The two hydrogen atoms fit into the two holes, acting as plugs for these holes; hence the two hydrogen atoms should be separated by an angle of 90 degrees, seen from the center of the oxygen atoms. The positive charges on the protons of the hydrogen atoms, however, repel each other and so increase the angle to slightly more than 90 degrees—in fact, 108 degrees (figure 35b). This is a typical plug-and-hole bond.

Another important molecule is carbon dioxide, which con-

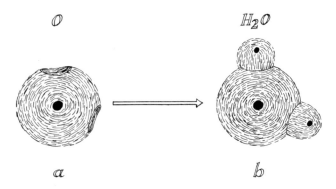

Figure 35
(a) The oxygen atom. The electron pattern has two holes at right angles from the nucleus. (b) The water molecule. The two holes in oxygen are plugged with hydrogen electrons. The hydrogen nuclei are within the hydrogen electron patterns. The small black circles are hydrogen nuclei. The large black circle is an oxygen nucleus.

sists of one carbon atom and two oxygens. This molecule also can be interpreted as a plug-and-hole bond. We again envisage the carbon atom as having four prongs. Here we may say that the four prongs are slightly bent; two go as plugs into the holes of one oxygen, the other two into the second oxygen. The result is a stretched structure with the carbon atom flanked by two oxygen atoms. (See figure 36.)

The carbon atom, with its four electron prongs, can form an unending series of molecules. The possibility of variety explains why carbon compounds are so common on Earth and play a central role in living matter. Let us look at a few of these structures. The simplest is methane (figure 37), with one hydrogen at each prong. We also can build a molecule with two carbons and six hydrogens; here all bonds are twin-electron bonds. The four prongs of carbon form twin bonds either with the electron of a hydrogen atom or with a prong of another adjacent carbon atom. The molecule is ethane. This principle can be continued, as indicated in figure 37, and we get a series of molecules called hydrocarbons: propane, with three carbons; butane, with four carbons; etc. The chain-

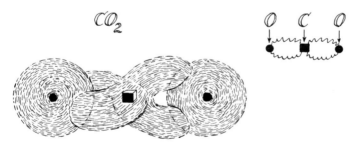

Figure 36
The carbon dioxide molecule $CO_2$. The four prongs of the carbon are plugging the holes of the two oxygen atoms. The schematic figure indicates the plug-and-hole bonds by wavy lines.

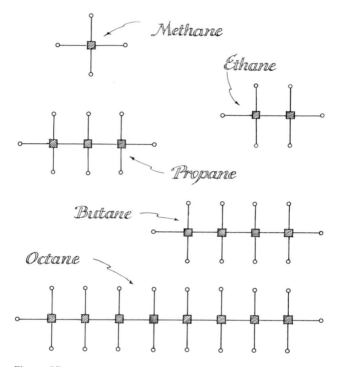

Figure 37
Hydrocarbons. Schematic pictures of the molecules. Squares are carbon atoms. Small circles are hydrogen. Straight connecting lines are twin-electron bonds.

like structures exist in any length desired. The short ones are
gases, the longer ones liquids, and the very long ones, solids.
They serve as burning fuel in the form of gas, oil, and candle
wax, and we shall see later why they are well fitted for burning.

The hydrocarbon chains are also important for our nutri-
tion when they end up with a characteristic arrangement of
atoms called the carboxyl group (figure 38). Such chains are
fatty acids, the constituents of animal fat.

Other characteristic carbon structures are the alcohol molecules shown in figure 39. Here the bonds toward the oxygen
are plug-and-hole bonds.

Another important group of long-chain molecules are the
carbohydrates. The chains are similar to the hydrocarbon
chains, but there is an oxygen added to each step of the chain.
The oxygen, as always, is tied on with plug-and-hole bonds.
The simplest carbohydrate is glucose, which is a form of sugar

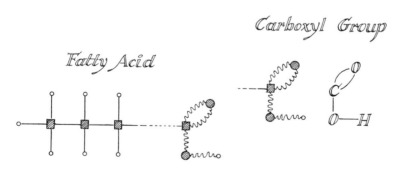

Figure 38
Fatty acid molecule. A long hydrocarbon with a carboxyl group at the right
end. The carboxyl group, COOH, consists of one carbon atom, two oxygens,
and one hydrogen bound together by plug-and-hole bonds. Squares are carbons; large circles oxygens; small circles hydrogens. Straight lines are twin-
electron bonds; wavy lines plug-and-hole bonds.

## Alcohol

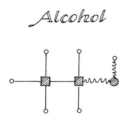

Figure 39
The alcohol molecule $C_2H_5OH$. Squares are carbons; large circles oxygen; small circles hydrogen. Straight lines twin-electron bonds; wavy lines plug-and-hole bonds.

## Sugar

Figure 40
Sugar. The glucose molecule $C_6H_{12}O_6$.

(figure 40). Cellulose is a very long carbohydrate. It is the material from which wood and other plant structures are made.

Next we come to a most important group of molecules, the amino acids. They are the building blocks of most living matter. Figure 41 shows the general principle of their structures. Carbon again, with its versatile four electron prongs, is the backbone of these molecules. The most characteristic, however, are the two end groups. At one end (on the right in the figure) there is a carboxyl group, and at the other end there is the amino group, $NH_2$. Between the two ends many different groups are found; each amino acid has its characteristic middle part. Figure 41 shows two of the simplest amino acids, glycine and alanine, and the general structure of the more complicated ones. The two end groups have a characteristic property: They can easily join. The amino group and the carboxyl group hang together (the so-called "peptide bond") so that amino acids readily form long chains, one amino acid hitched onto the next. Such chains are called proteins; they play an important role in the functioning of life, as we shall learn in chapter 8.

We now come to the third kind of chemical bond, the *ionic* bond. It occurs between a pair of different atoms; one atom must possess a very loosely bound electron, which, therefore, can easily be removed. Once removed, the atom becomes

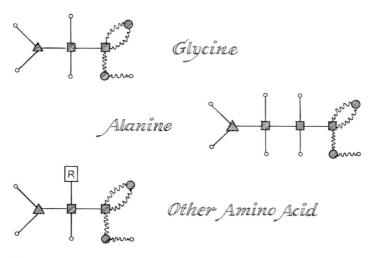

Figure 41
Amino acids. Amino acids have the carboxyl group at one end and the amino group $NH_2$ at the other end. The symbol $R$ stands for various kinds of chemical compounds.

positively charged; such an atom is called an *ion,* specifically a *positive ion.* Examples of such atoms are the lithium and sodium atoms. In either case, relatively little energy is needed to remove an electron. The other partner in an ionic bond must have the opposite property: It must welcome an additional electron into its electron configuration. By this, we mean that the atoms should attract and bind an additional electron; energy would be needed to remove it. Such an atom with a surplus electron is called a *negative ion,* since it would be negatively charged. Chlorine and fluorine are good examples of atoms in which the electron configuration has a hole where an additional electron easily fits.

There is a strong electric attraction between a positive and a negative ion. Therefore such a pair forms a molecule with no total charge but with a strong bond, since it needs a good amount of energy to separate the two ions of opposite charge. Example are the molecules LiF (lithium fluoride) and NaCl (sodium chloride, the molecule of rock salt).(See figure 42.)

## Crystals and Metals

So far we have described a few examples of molecules in which a definite number of atoms bound together. The hydrogen molecule $H_2$ consists of two hydrogen atoms, the water molecule consists of one oxygen atom and two hydrogen atoms, and so on. But there also exist structures that consist of a very large number of atoms of one or two kinds that are bound together by the chemical bonds. These structures are accumulations of an enormous number of atoms of the same kind; indeed they are so big that they appear to us as chunks of solid matter. The number of atoms is not fixed; any number of atoms can be added or removed, changing only the size but not the intrinsic nature of the substance. We shall discuss four examples of such structures: a diamond crystal, a quartz crystal, a metal, and a salt crystal. Diamond consists of

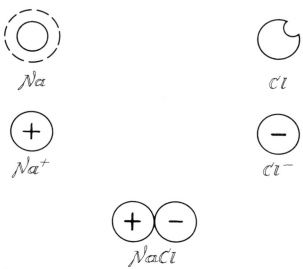

Figure 42
Ionic molecules. First row shows the atoms of Na and Cl. The Na has a weakly bound outer electron; Cl has a hole in the electron configuration. Second row shows the ions. The weakly bound electron of Na is removed, forming a positive ion Na⁺. The hole is plugged by an electron, forming Cl⁻. Third row shows the molecule NaCl.

carbon atoms. Remember that each carbon atom possesses four electron prongs in the directions of the corners of a tetrahedron. It is easy to see in figure 43 that carbon atoms can form an extended three-dimensional net by repeated joining of two prongs from neighboring atoms into an electron-pair bond. This structure can go on indefinitely in all directions. This indeed is the internal structure of diamond. The hardness of that material comes from the excellent overlap of the prongs, which form very good and strong bonds. A diamond crystal of, say, one millimeter size contains about $10^{21}$ carbon atoms; we may consider it as one giant molecule.

A quartz crystal is similarly constructed. It is made of silicon

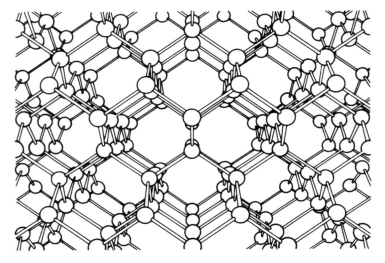

Figure 43
Diamond lattice. Carbon atoms are bound together by four bonds emerging from each atom. There are two bonds per atom, since each bond belongs to two atoms.

and oxygen atoms in the ratio one silicon to two oxygen. The silicon atom is similar to carbon; it also has four electron prongs, but not as strongly developed. In a quartz crystal each prong of a silicon atom forms a plug-and-hole bond with an oxygen atom. Remember, oxygen is an atom with two holes in its electron configuration. The prong of one silicon fits into one hole and the prong of a neighboring silicon fits into the other. We then get an arrangement of silicon atoms not unlike the one of carbon atoms in diamond, except that there is always an oxygen placed between neighboring silicons. (See figure 44.) Again, this structure is rather stiff and hard. Quartz and similar structures make up the rocks of our mountain ranges.

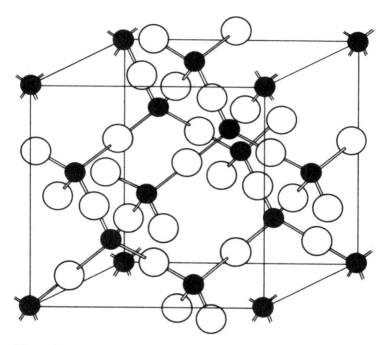

Figure 44
Silicon oxide lattice. Four bonds emerge from each silicon atom (small, black circles) and two from each oxygen (large, white circles).

The structure of a metal can be regarded as a vast extension of the electron-pair bond, such as the one in the hydrogen molecule $H_2$. How does it work? Metals are formed by atoms in which the last few electrons (one, two, or three) are very loosely bound. When such atoms are brought close together, the loose electrons no longer stay with the atom to which they belonged originally, but move more or less freely through the body of the material. This is why metals are such good conductors of electricity. The loose electrons form what is called an electron gas within the metal. The Pauli principle does not allow more than two of them to be in the same quantum state of motion; still the electrons move as a gas and, altogether, have a lower energy than they would have if they still were attached separately to their atoms. Why? Because, as a gas, they roam around freely within the metal and therefore get attracted by the nuclei of several atoms. The lowered energy is what makes the metallic bond; energy must then be supplied to liberate an atom from the metal. This energy is quite high but somewhat less than the bond energy in diamond or quartz; thus, metals are softer, and melt at a lower temperature.

Our last example of a solid representing large molecules is a rock-salt crystal. We described before the ionic bond between two types of atoms, one which forms a positive, the other a negative, ion. This bond can easily be extended to an arbitrarily large number of such pairs. Then the ions are placed in the manner of a three-dimensional chess board; a positive ion always has negative ions as neighbors, and vice versa. The rock-salt crystal is such a structure; the positive sodium ions ($Na^+$) alternate with negative chlorine ions ($Cl^-$), as sketched in figure 45. The electric attraction between positively and negatively charged neighbors provides its rigidity and strength.

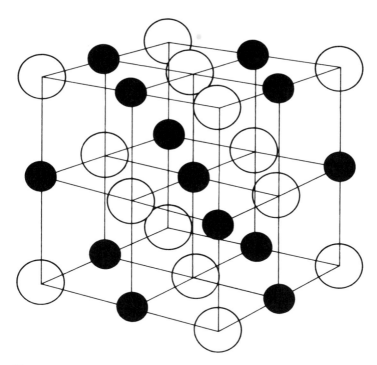

Figure 45
Rock-salt crystal lattice. Positive sodium ions (small, black circles) alternate
with negative chlorine ions (large, white circles).

## Chemical Energy, Chemical Burning

So far, we have given a descriptive account of a few of the most common molecules, including some of the solids formed by the chemical bond. Conditions in our environment on Earth are such that these structures are constantly formed and decomposed. If the earth were much hotter, as on the surface of the sun, molecules would never form because the thermal agitation would be too strong—the atoms would not stay together. If it were much colder, the atoms within the molecules and crystals would remain bound and no changes would occur. The temperatures on Earth are such that enough energy is available to occasionally break up some molecules, yet compounds can exist for some time. The making and breaking of these structures characterizes our environment, giving it the ever-changing variety we see and establishing the conditions for the existence of life.

One of the most important consequences of formation of molecules is the release of energy. This release is most clearly seen when we burn coal or other substances. Every burning is connected with the formation of new molecules, and that is where the heat energy comes from. We must now see more clearly how and why energy is gained when atoms join to form molecules. A chemical bond represents energy because a certain amount of energy is needed to break a bond; hence the same amount is gained if the bond is formed. As a simple example of a nonchemical bond, let us consider a magnet holding a piece of iron at its poles by magnetic attraction (figure 46). It takes a certain amount of energy to remove the iron from the magnet. When the piece of iron is returned to the magnet, the same amount of energy is gained; the pull from the magnet has produced it. When we remove the iron from the magnet, our muscles supply the energy needed. Where does the energy appear that is produced when the

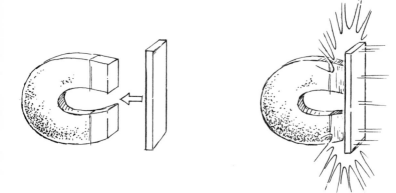

**Figure 46**
When the iron hits the magnet, the energy of magnetic attraction is transformed into heat and sound.

piece is returned to the magnet? The energy appears partly in the form of sound, when the iron slaps on the magnet, and partly in the form of heat—the iron is heated when it hits the magnet with force. The energy could even be used mechanically if the piece of iron were rigged with strings and pulleys to perform work when attracted toward the magnet. (See figure 47.)

Similar exchanges of energy occur in the case of the chemical bond. It takes energy to separate a molecule into atoms, and energy is gained when the atoms form a molecule. The energy gained turns up in many forms. For example, it appears in the form of vibrations. When atoms slap together, the molecule formed is set in vibration by the energetic encounter. Energy also appears in other forms of motion. The energy produced when the atoms collide and enmesh is then transferred to other neighboring molecules, whose motion is

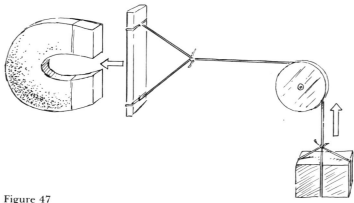

Figure 47
In this arrangement, the energy of attraction is used to lift a weight.

thereby accelerated. Altogether, whenever atoms form molecules, energy is released and usually turns up in the form of general motion, which is equivalent to heat.

Some chemical bonds are stronger, others weaker. The bond between the two H-atoms in $H_2$ is worth 4.6 eV; that means 4.6 eV of energy is necessary to separate the molecule $H_2$ into two hydrogen atoms. The bond between *one* H-atom and the oxygen atom in water ($H_2O$) is equal to 4.8 eV. Since the water molecule contains two H-atoms, one needs $2 \times 4.8 = 9.6$ eV to separate a water molecule into its three atoms, H, H, and O. In the molecule $CO_2$ the total energy required to separate it into one C- and two O-atoms is 16.7 eV. Since there are four plug-and-hole bonds in $CO_2$, the energy per bond will be 4.2 eV. In all three examples the energy per bond is a little more than 4 eV. The binding energy in crystals and metals is of the same order of magnitude. It takes about 9.4 eV to release a carbon atom from a diamond crystal. Only a little less, about 8.6 eV, is necessary to get a carbon atom out of a piece of coal. This again represents 4.7 eV, or 4.3 eV, per bond, since each carbon atom is bound by two bonds, both in the diamond crystal and in the coal.

The bond between the two oxygen atoms in the oxygen molecule $O_2$ is a weaker bond, however. After all, the two holes in each oxygen atom do not lend themselves easily to bond formation. Indeed the energy of the bond in $O_2$ is only 5.2 eV—that is, 2.6 eV per bond, since we consider the connection of one hole in one atom with one hole in the other atom as one single bond. The binding energy in a metal is also a weaker bond. In copper, for example, it is only 3 eV per atom.

Let us consider in more detail a chemical process in which energy is released: the burning of hydrogen gas. What happens here? Burning occurs when oxygen is present, as it always is in air. In the burning of hydrogen, two hydrogen molecules and one oxygen molecule form two water molecules. As shown in figure 48, the process goes in two steps. First, the hydrogen molecules and the oxygen molecule are decomposed into atoms. This step needs energy: $2 \times 4.6$ eV for the two hydrogen molecules and 5.2 eV for the oxygen makes 14.4 eV altogether. Then with the formation of two

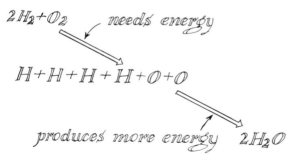

Figure 48
The burning of hydrogen.

water molecules, we gain $2 \times 9.6 = 19.2$ eV. Thus we have a net gain of $19.2 - 14.4 = 4.8$ eV per process; quite a lot of energy. It shows up in the form of heat motion or light emission. We will see a bright flame. The reason for this gain is that the break up of the oxygen molecule needs only 2.6 eV *per bond* whereas the bonds in water represent 4.8 eV. So, whenever an oxygen atom is transferred from an oxygen molecule into another molecule (water in this case) it goes from a state of weaker bondage (oxygen molecule) into a state of stronger bondage (water molecule), and that leads to a gain in energy.

The burning process takes place in two steps. The first is the splitting of the molecule to be burned (hydrogen) and the oxygen molecule, and the second is the joining of the atoms to form the final molecules (water). The heat energy at normal temperature is not high enough to split the molecules into atoms. Combustion does not take place if it is not kindled. But if we provide heat from the outside in the form of a lighted match or kindling wood, the reaction starts. We need only to provide heat for the starting of the process because the second stage, the formation of $H_2O$, supplies more energy than required for the subsequent splitting of $H_2$ and $O_2$. The burning of $H_2$, therefore, produces energy in the form of heat, even though part of the energy produced in the second stage must be used to initiate further reactions. The useful heat is the relation of the surplus of the energy in the second stage to the energy needed in the first.

Once started, the production of water proceeds until all the hydrogen is used up. The heat produced is so great that it causes atoms and molecules to emit light. The flames we see leaping up are molecules ejected by the intense heat and made to radiate their characteristic light. Fire, therefore, is not a new form of matter, as people once believed; it is the material made incandescent by the large amount of energy set free in a chemical reaction.

The burning of hydrogen is the simplest form of chemical energy production. The burning of coal is not very different. The reaction is $C + 2O = CO_2$; a carbon and two oxygen atoms form carbon dioxide. A carbon atom must be released from coal, which requires 8.6 eV. An oxygen molecule in the air must be separated into two oxygen atoms, requiring 5.2 eV. The formation of $CO_2$ gains 16.7 eV. Thus the net gain of burning coal is $(16.7 - 8.6) - 5.2 = 2.9$ eV per atom burned, which is equivalent to 6.6 kilowatt hours per kilogram of coal. The flames that we see when coal is burning are molecules of $CO_2$ and small particles of coal ejected by the heat energy produced in the burning reaction. The heat makes them radiate light of the characteristic color of the flames.

A similar process occurs in the burning of methane (ordinary cooking gas) or any other hydrocarbon. Here again, initial heat is necessary to split not only the $O_2$ molecules in air but also the hydrocarbon molecules. Then the carbon combines with oxygen to form carbon dioxide and the hydrogen combines with oxygen to form water. The chemical reaction for the burning of methane is shown in figure 49. We get a net energy yield because the break up of oxygen molecules into oxygen atoms needs less energy per bond than there is gained when the oxygen forms carbon dioxide or water. Therefore methane and the other hydrocarbons burn in air with intense flame and heat. In the case of methane burning, however, both $H_2O$ and $CO_2$ are produced. The flame contains not only incandescent $CO_2$ gas but also water vapor. If you hold a cold piece of shiny metal in a candle or gas flame, you will find condensed water vapor on it.

Chemical reactions that produce energy must always start with molecules that have weaker bonds and end with mole-

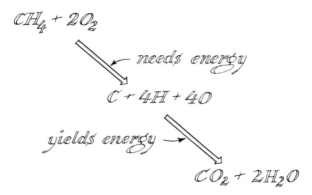

Figure 49
The burning of methane.

cules that have stronger bonds. The difference between the weaker and stronger bond strengths is released as energy. In most cases of burning, the weak bonds are the bonds in the oxygen molecule $O_2$, which, in the burning process, are split and form molecules such as $CO_2$ and $H_2O$ in which they are more strongly bound.

**The States of Aggregation**
The chemical bond keeps the atoms tied within the molecules and also within metals and certain crystals such as diamond, quartz or rock salt. Are there also forces between molecules? There must be such forces, since we know, for example, that water molecules form ice and hold together in water. The forces between molecules have quite different origins than the chemical force. Let us take water as an example. The water molecule is more positively charged on the side where the hydrogen atoms sit and more negatively charged on the opposite side. This gives rise to a weak bond between water

molecules when they are arranged such that the positive side of one molecule is near the negative side of another. This bonding effect is called the *hydrogen bond*. Do not confuse this with the very much stronger twin-electron bond that holds two hydrogen atoms together in the hydrogen molecule $H_2$.

In cases where there are no positive and negative charges situated on the molecules, there still can be causes for intermolecular attraction. For example, when two molecules come near each other, the electron patterns in each unit have a tendency to vibrate in unison. This tendency produces an attraction, the so-called *van der Waals force*. Its strength depends on the nature of the electron patterns and is different in different molecules. But it always is an attractive force, much weaker than the chemical bond.

These weak intermolecular forces keep molecules together in the form of liquids or solids, but at much lower temperatures than the solids and liquids that are held together by chemical forces, such as metals or rocks. In water and ice, for example, the hydrogen bond provides the bonding, but it is too weak to keep the molecules together above 100° Celsius, when water turns to vapor. Oxygen molecules ($O_2$) are held together by van der Waals forces. They are so weak that a solid is formed only below $-183°C$. It is a liquid up to $-110°C$, but above this temperature the forces can no longer keep the oxygen molecules together; it becomes the gas that is part of our air.

Why do molecules and atoms sometimes aggregate as liquids, sometimes as solids, and sometimes, when they form gases, not at all? At low temperatures almost all substances become solid; at high temperatures all substances are gases; and at some intermediate temperatures they all are liquid. The values of the temperatures at which these transitions occur (melting point, boiling point) differ very much for different substances. These values depend on the strength of the attraction between the constituents.

At very low temperatures there is very little heat motion, and therefore the atoms or molecules can arrange themselves neatly in regular arrays and are kept in these patterns by the forces between them. They form a solid. The regular arrangement of the molecules is often beautifully displayed in crystals. The field-ion microscope displays such a regular pattern in tungsten, as we saw in figure 20. Crystals are solids in which the ordered pattern extends over a large volume and becomes directly visible in the straight edges and geometrically arranged planes and points. Since they multiply the tiny order by an enormous factor, crystals give us a direct picture of the symmetry of the smallest units. If you try to deform a solid by bending or breaking, you feel a resistance—you are changing the arrangement of the constituents and must overcome the forces that keep them in the regular pattern.

At higher temperatures the irregular heat motion becomes stronger and destroys the ordered arrangement of molecules or atoms in solids. Still the forces between the constituents keep them tightly packed adjacent to each other, but they no longer can force them into a regular array. There will be no resistance to deforming or bending, but the constituents will stay together. They form a liquid. The transition from the solid to the liquid depends on the strength of the forces between the constituents. If the forces are strong, as in rock or metal, it will take a high temperature to overcome the bonds that keep them in the regular pattern. If they are weak, as in water or air, the transition occurs at a lower temperature.

If the temperature rises still higher than the melting point, the heat motion becomes stronger, until it overcomes not only the ordering effect but also the packing effect of the forces.

Then the atoms or molecules torn from each other by the heat motion fly off in all directions, colliding with each other and hitting the walls of the container. They no longer are tightly packed but are spread throughout the entire container. They form a gas. The point at which a gas is formed depends again on the strength of the forces. In oxygen the forces between the molecules are so weak that they cannot keep the molecules together even at ordinary temperatures. In some metals and rocks the forces are so strong that a temperature of several thousand degrees centigrade is needed to produce the gaseous state.

The increase of temperature reduces the specific properties and the organization of the substances. In the solid state, substances exhibit such typical shapes as crystals, have a specific texture and hardness, and are easily identifiable. In the liquid state, shape and texture are gone; the substance assumes the form of the container. Only a characteristic density, color, and surface remain. This transition is displayed dramatically when a snowflake, with all its intricate hexagonal structure, melts into a shapeless drop of water. The gaseous state has even less specificity. There is no characteristic density any more, no surface, nothing but a shade of color and smell. Still, in all three states, the substance consists of the same molecules or atoms. The same metal atoms form solid or liquid metal or gaseous metal vapor. The same water molecules form ice, water, and steam.

# 7 The Quantum Ladder

## Size and Stability

We found in the previous chapter that everything we see around us is a combination of ninety-two atomic nuclei and their electrons. The substances and the shapes of all things are the result of an interplay of various electron patterns and their combinations. Electrons assume their characteristic wave patterns when they assemble around the atomic nuclei; they form atoms and the atoms join to form molecules. These patterns are responsible for the specific properties of all materials and give each substance its special character.

The patterns can be deformed and destroyed in energetic collisions or other strong perturbations, but they exhibit a certain typical stability. We have seen, for example, that at the temperatures we experience here on the earth's surface the heat motion is not energetic enough to destroy atomic and most molecular patterns. This is the reason the substances we find in our immediate surroundings have specific properties. The limit of stability of an atom or molecule—that is, the energy necessary to change a specific pattern—is different for the various types of atoms and molecules. It depends on many factors, but mainly on the size. Typically, large units have lower stability than small ones. Big molecules can be broken up more easily than small ones. The extremely large molecules that make up organic material, such as meat or vegetables, are broken into pieces in the process of digestion, which goes on at a relatively low temperature in our stomach. But as we saw in the last chapter, the breakup of the much smaller oxygen molecule needs the temperature of a flame. The removal of an electron from a hydrogen atom, the smallest of all atoms, needs even higher energies; it cannot be done in a flame, but only in a strong electric discharge.

The relation between size and stability is a direct consequence of the wave nature of particles. Let us remember the fact that a long piano string gives a lower pitch and lower overtones than does a short string. In close analogy, a large wave pattern would have lower frequencies, too. According to our fundamental frequency-energy relation, lower frequency means lower energy. We expect a wave pattern of large size to possess less energy and also to be sensitive to perturbations of lower energy. This is why the size-stability relation has such general validity. The smaller the system, the higher will be its stability and the more energy will be needed to change its characteristic structure.

**The Structure of Nuclei**
As we have seen, matter consists of electrons and atomic nuclei. The nuclei are different in each element. They differ in mass. For example, hydrogen nuclei are lighter than oxygen nuclei, which in turn are lighter than iron nuclei. But the most important property that characterizes the nuclei of the different elements is their electric charge. This charge is responsible for the character of the atom, since it determines the number of electrons and therefore the patterns that the electrons assume. The number of electrons is equal to the number of charge units on the nucleus, a number we earlier called the atomic number Z. At this stage of our story the nucleus of an atom is an indivisible entity, whose charge and mass are characteristic of each type of atom. Each element has its own typical nucleus. Matter is made up of many different "elementary particles," the electrons and the various atomic nuclei, a different one for each element.

But it is not a satisfactory state of affairs to consider each of the different nuclei of the ninety-two elements as a different elementary particle. One would prefer to think that the various types of atomic nuclei are all combinations of a few simple constituents. Any nucleus would be a structure made up of these constituents, rather than the nucleus itself being an elementary particle.

So far, we have considered the atomic nuclei as massive particles, endowed with a positive charge but seemingly without any internal structure. Could it be that their lack of structure is apparent only, and that their internal structure, just as in the case of atoms, shows up only above an excitation threshold? The small size of the nucleus would indicate a very high threshold according to our size-stability relation, very much higher than the corresponding energies in atoms. Perhaps we should expect that the internal structure of the nucleus would be of no importance in the dynamics of atoms, and would be observed only when much higher energies came into play than those with which we deal in atomic or molecular problems.

It was one of the most remarkable developments in modern physics when experiments revealed the same kind of quantum world within the tiny nucleus as we have found in the much larger atom. Atomic nuclei indeed have a structure, and they are made of two kinds of particles, protons and neutrons. The proton is identical to the lightest atomic nucleus, the nucleus of the hydrogen atom; it carries one unit of positive charge, and its mass is 1,836 times heavier than the electron. The neutron is a particle of almost exactly the same mass as the proton, but without any charge.

Since each proton carries one unit of charge, the charge of the nucleus is equal to the number of protons contained in the nucleus. Thus the number of protons in each nucleus is the same as the characteristic atomic number Z of the element. The little table of Z-values in chapter 4 gives the number of protons contained in the nuclei of the elements. The neutrons have no charge; they only add to the weight of the nucleus.

The number of neutrons in a nucleus is either equal to or slightly greater than the number of protons.

There must be a force keeping the neutrons and protons together in so small a volume as the atomic nucleus represents; nuclei are about 10,000 times smaller than atoms. The nuclear force acts between the protons and neutrons and is able to confine them within the volume of the nucleus. It is a very strong attractive force; it not only holds many particles within the nuclear boundaries but also must overcome the electric repulsion between the positively charged protons.

This nuclear force is a new kind of fundamental force in nature, different from everything we knew before. Its discovery was an important step in the attempts to gain insight into the deepest structure of nature. Until then, we knew only of two kinds of fundamental forces, the gravitational and the electromagnetic forces. Gravitation governs the movements of big units of matter, the movements of celestial bodies, and the falling of objects on earth. On the atomic scale, however, gravitational forces are very, very weak; therefore they are not important inside atoms and molecules. The electric force acts between the atomic nuclei and the electrons and produces the special patterns of electron waves that are responsible for the structure of atoms and molecules. The nuclear force acts between the protons and the neutrons and holds them together in the atomic nucleus.

We have no direct human experience of the nuclear force, but everybody has felt the force of gravity on a heavy object and has seen the effect of electric or magnetic forces. Gravity and electricity have a long range. Even though the effect of gravitational and electric attraction decreases with increasing distance in accordance with the well-known inverse square law, in principle it is noticeable at any distance. The nuclear forces are much harder to detect because of their very short range. Nuclear forces break off completely at very short distances. They have been found to act over stretches of only $10^{-13}$ centimeters, a length that is 100,000 times smaller than the size of an atom. Of course it is impossible to notice directly the effect of nuclear forces on an ordinary human scale.

Indirectly the nuclear forces have an enormous effect; if these forces did not exist, the nucleons (nuclear particles) would fly away from one another. Without them there would be no atomic nucleus, except a proton; hence there would be no atoms except hydrogen. Furthermore, as we shall see later, the solar energy supplying us with heat and power comes ultimately from the effects of nuclear forces, and recently man has been able to use and abuse the effects of nuclear fission, which are a direct consequence of nuclear forces.

Nuclear physics proceeded along the same lines as atomic physics. In the 1930s, only ten years after the discovery of the wave nature of atomic electrons, specific quantum states were found within the nucleus. The nuclear force confines the motions of protons and neutrons to a small region—the volume of the nucleus. This confinement is similar to the confinement of electrons within the atom by the electric attraction of the atomic nucleus. The confining effect of the nuclear forces should produce patterns of proton and neutron waves similar to the electron patterns in atoms. It is true that the wave properties depend on the mass of the moving particle: the heavier the particle, the shorter the wavelength and the harder it is to observe wave effects. Still the wave effects give rise to characteristic wave patterns as soon as the motion of the particles is confined.

Thus we find the same quantum effects repeated again within the nucleus that we have observed in the atom. Quantum states, stability thresholds, characteristic patterns, identity of nuclei of the same type, all are observed again on a much smaller scale in size, but, because of the size-energy relations, on a much larger scale in energy.

In order to find out about the structure of the atomic nuclei one must overcome the threshold energies of the nuclear quantum states. These energies were found to be as high as a hundred thousand to a million electron volts. It is hard to concentrate this amount of energy on a single nucleus. Flame or electric discharge will no longer do the job, as they did for atoms. The first experiments of "smashing" an atomic nucleus were made with alpha particles, which are ejected with very high energy from certain radioactive substances. These particles are identical with the nuclei of helium. They consist of two protons and two neutrons in a tightly bound unit. They are the same particles with which Rutherford discovered the existence of the atomic nucleus in 1911. Only a few years later, in 1919, Rutherford again used alpha particles for another fundamental discovery. When he directed alpha particles into nitrogen gas, he found that the bombardment could break the nitrogen nucleus into pieces. He was able to show that a proton was broken off from the nitrogen nucleus, and thus he proved that nuclei contain protons as constituents.

Since this memorable date an enormous amount of knowledge has been accumulated in regard to the structure of atomic nuclei. After 1930 machines were invented and constructed to accelerate protons or alpha particles artificially to very high energies; so we no longer need to use radioactive rays for the investigation of atomic nuclei. These machines have different names: cyclotrons, synchrotrons, electrostatic accelerators, etc. Popularly they are called "atom smashers," but they should be called "atomic nucleus smashers." The atom is easily taken apart by heat or electrical discharges.

When we light a match, we take atoms apart; many atoms in the tip of the match lose one of their electrons in the little explosion that occurs. But it is the atomic nucleus, not the atom, that resists any interference with the structure until one reaches energies of millions of electron volts.

In a way, however, the word "atom smasher" is apt. The atomic nucleus is not only the center of the atom, it is the essential part. Almost all the mass of the atom resides in the nucleus. The surrounding electrons weigh less than $1/2.000$ of the total mass. They can be removed and replaced, but it is the charge of the nucleus that determines the pattern in which the electrons surround the nucleus, and it is this pattern that determines the properties of the atom. Therefore the nucleus is the part that characterizes the atom with its special properties.

When Rutherford and his collaborators for the first time changed one atomic nucleus into another, they fulfilled the great dream of the alchemists, the transmutation of one element into a different one. Rutherford bombarded nitrogen with alpha particles, which carry a charge of two units. An alpha particle penetrated into a nitrogen nucleus and broke off one proton; thus two charges were added and one (the proton charge) was taken off. The total charge on the nucleus was raised by one unit. This gain transformed the nitrogen nucleus into an oxygen nucleus. The alchemists' particular fantasy of transmuting lead into gold was achieved much later with a similar method. Of course the energy cost of doing this commercially is prohibitive.

As mentioned earlier, nuclear physics showed that, like the electrons in atoms, protons and neutrons form wave patterns in the nucleus leading to specific quantum states. (See figure 50.) When changing from a higher energy quantum state to a lower one, the nucleus, as the atom does, releases the energy difference in the form of a light quantum whose frequency $\nu$ corresponds to the energy difference $E$, according to Planck's formula $E = h\nu$. The nuclear case differs from the atomic case in the amount of the energies involved. Atoms usually emit visible light; atomic nuclei emit light of much higher frequency because the energy differences are more than a hundred thousand times larger. This light is like very penetrating X rays; it is referred to as "gamma rays."

It is also useful to compare the characteristic time intervals of the dynamics of nuclei and atoms. We have referred to an "atomic year" as the approximate time it takes an electron to circle around the nucleus; we found it to be $10^{-16}$ seconds. The time of revolution of a nuclear particle within a nucleus—the "nuclear year"—is $10^{-23}$ seconds. Whether a process is considered to be "fast" or "slow" in these domains is determined by comparing their duration with that of the appropriate "year."

There is also another difference between the quantum mechanics of atoms and of atomic nuclei. In the atomic case we know exactly the nature of the force confining the electrons to the nucleus; the electrostatic attraction. It is possible to calculate the electron wave patterns and to predict accurately the energy and shape of the quantum states. In the case of the atomic nucleus, the confining force is new to us; it is more complicated than the electrical force and not well understood. As a result, we cannot predict the wave patterns as well as we wish. All we can do at present is try to determine the properties of the new force by studying its experimental consequences. You can appreciate the difficulties when you consider that atomic nuclei are ten thousand times smaller than atoms. Still, a considerable knowledge has been accumulated regarding the new force. We know at present its inten-

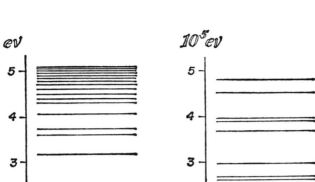

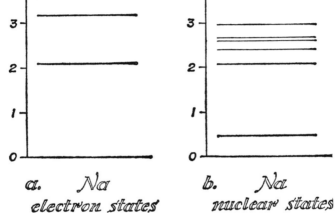

Figure 50
The quantum states of the sodium electrons (a) compared with the quantum
states of the sodium nucleus (b). The states (a) are the same as in figure 25 in
chapter 5. Note that the scale of (b) is 100,000 times larger.

sity, its range, and some of its detailed properties, including the remarkable fact that, although attractive over most of its range, the force turns repulsive if the nucleons come very close to each other.

Nuclear physics has taught us one outstanding lesson. All matter consists of three types of elementary particles— protons, neutrons, and electrons. Everything we see about us is a combination of these three entities. Protons and neutrons combine to form atomic nuclei; electrons fall into their patterns around the nucleus and form the atom; atoms combine into molecules and molecules or atoms aggregate into matter as we see it around us. A great step has been achieved by this reduction of the variety of substances in our environment to only three elementary units, whose various combinations under the influence of nuclear and electromagnetic forces make up most of the materials of the universe.

Even protons, neutrons, and electrons may not be elementary. Indeed, modern subnuclear research has shown that protons and neutrons probably consist of subparticles called *quarks*, which we will discuss later in this chapter.

### Isotopes, Radioactivity

An atomic nucleus is composed of neutrons and protons held together by the nuclear forces. The proton is charged, the neutron is not; hence the charge of the nucleus is determined by the number of protons. This charge is an important quantity, since it determines the kind of atom that will be formed around the nucleus, and therefore characterizes the element to which the nucleus belongs. The neutrons serve only as fillers, which increase the mass of the nucleus.

The nuclear forces lead to the greatest stability when the number of neutrons is about equal to or slightly larger than the number of protons. This is the arrangement we find in most nuclei. For example, the helium nucleus, with a charge of two units, consists of two protons and two neutrons; the carbon nucleus has six protons and six neutrons; nitrogen has seven protons and seven neutrons; fluorine has nine protons and ten neutrons.

Sometimes a certain number of protons form nuclei with different numbers of neutrons. The different nuclei are of the same element (the element is determined by the number of protons) but of different weight. Two such samples of the same element that differ only in the number of neutrons in their nuclei, and hence in atomic weight, are called *isotopes*. For example, a carbon nucleus with six protons and seven neutrons is an isotope of ordinary carbon, which has six neutrons in its nucleus. Also, there is a nitrogen nucleus with seven protons and eight neutrons, an isotope of the ordinary nitrogen with seven neutrons. We denote the two carbon isotopes with the symbols $C^{12}$ and $C^{13}$ and the two nitrogen isotopes with $N^{14}$ and $N^{15}$. The number indicates the total number of constituents, neutrons and protons together. The isotopes $C^{13}$ and $N^{15}$ are much rarer than the ordinary types of $C^{12}$ and $N^{14}$.

What about further isotopes, such as $C^{14}$ (eight neutrons) or $C^{11}$ (five neutrons)? Such isotopes are indeed possible. Isotopes $C^{11}$ and $C^{14}$ can be formed by suitable application of accelerator machines, but such nuclei with an abnormal surplus of one kind of particle exhibit a strange phenomenon. This phenomenon, which occurs whenever the number of neutrons gets much out of balance with the number of protons, is called *radioactivity*. Slowly but surely, a proton transforms itself into a neutron when there are too many protons, as in $C^{11}$; or a neutron changes into a proton if there is an abnormal surplus of neutrons, as in $C^{14}$. Then $C^{11}$ becomes a nucleus with six neutrons and five protons, and that is a boron nucleus, $B^{11}$.

The isotope $C^{14}$ transforms into a nucleus with seven protons and seven neutrons, and that is a nitrogen nucleus, $N^{14}$.

This transformation process is of special interest. It occurs steadily and slowly with a fixed half life, which is twenty minutes in the case of $C^{11}$ and 4,700 years in the case of $C^{14}$. We use the term "half life" because after twenty minutes half of the $C^{11}$ nuclei become $B^{11}$ nuclei; after another twenty minutes half of the remaining $C^{11}$ nuclei are transformed, etc. The same regular process occurs with $C^{14}$ in steps of 4,700 years.

The slow transformation of a proton into a neutron inside a nucleus and the reverse transformation of a neutron into a proton is characteristic of the nuclear world. There is nothing exactly like it in the atomic world, where electrons may change from one energy level to another but remain electrons. The *consequences* of the nuclear transformations are even more striking. Indeed, some of these consequences are so bizarre that only the accumulation of several independent lines of evidence over decades has convinced us that they must take place. In spite of this, they follow some strikingly simple rules, such as the laws of conservation of energy and of charge. The same laws were fulfilled in atomic transitions.

There are some other points of similarity between those bizarre nuclear processes and the familiar atomic transitions. When the electron configuration in an atom undergoes a transition from a higher energy state to a lower one, the excess energy must "go somewhere" (conservation of energy). For an isolated atom this excess energy is given off in the form of a quantum of light, a *photon*. The photon is created at the moment of transition; it did not exist in the atom before the transition.

In some ways, nuclear transformations are similar to atomic transitions; in other ways they are radically different. When a neutron changes into a proton in the transformation $C^{14} \rightarrow N^{14}$, conservation of charge requires that the extra positive charge must come from somewhere. But for an isolated nucleus there is nowhere for the positive charge to come from. However, to receive a positive charge is equivalent to giving off a negative charge. And this is just what happens: In the transition $C^{14} \rightarrow N^{14}$ a negatively charged electron indeed is emitted. Giving off a negative electron while changing a neutron into a positive proton has the result that no net charge is created or destroyed; so the law of conservation of charge is satisfied.

Thus when a carbon $C^{14}$ nucleus changes into a nitrogen $N^{14}$ nucleus, an electron is given off. Where did the electron come from? Was it already "hiding" in the nucleus? No, it was created during the transformation of neutron into proton, just as light quantum is created during the transition of atomic electrons from a higher to a lower energy state. Hard to believe? Yes. In fact, people thought for years that the electron already existed in the nucleus. Later they convinced themselves that the nucleus cannot contain an electron, just as an atom cannot contain a photon.

Is this the end of the story of the transformation of $C^{14}$ into $N^{14}$, that a neutron is transformed into a proton and an electron given off? No, there is more. Not only must charge be conserved but energy must be conserved also, as in the case of atomic transitions. Surprisingly, in the nuclear transition the observed energy of the outgoing electron is much smaller than the energy change of the nucleus; some surplus energy is not accounted for. Therefore our description of the transformation must be incomplete. Something else must be given off in addition—something hard to detect, since no one had seen it before.

Indeed, something else *is* given off, something so hard to

detect that it was not observed for a long time after its existence was postulated in order to preserve the law of conservation of energy. This new particle is the *neutrino.* The neutrino carries no charge and therefore is not subject to electrical attraction or repulsion. The neutrino also has zero mass, like the light quantum. It travels through matter almost without hindrance and therefore is extraordinarily hard to detect. Its existence was postulated in 1930 to explain the kind of nuclear transformation described here, but only in 1956 was it detected beyond reasonable doubt. Yet long before this detection, its properties were predicted in detail on the basis of what was otherwise missing from nuclear transformations.

The story of the transformation $C^{14} \to N^{14}$ is now complete: A neutron is transformed into a proton and an electron and a neutrino are emitted from the nucleus. (See figure 51.)

What about the nuclear transformation $C^{11} \to B^{11}$, in which a proton changes into a neutron? Here a positive charge should appear in order that total charge not change. This positive charge was observed; it was a positive electron! The new particle was given the name *positron.* The positron is easy to detect because of its charge. It has the mass of the electron but a positive rather than a negative charge. And again, along with the positron, a neutrino is also emitted. Thus the transformation of carbon $C^{11}$ into boron $B^{11}$ involves the transformation of a proton into a neutron in the nucleus and the emission of a positron and a neutrino.

The existence of the positron—an electron of positive charge—is itself a tremendous story. It was the first example of *antimatter,* to which we shall return.

What is new in the study of nuclear transformations is the appearance of previously unknown particles—positrons and neutrinos—and the fact that pairs of particles are created, just as photons are created in atomic transitions. The particle

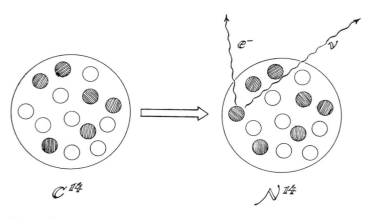

Figure 51
Radioactive transformation of $C^{14}$ to $N^{14}$. Open circles are neutrons, dark circles protons. One neutron to the extreme left of $C^{14}$ changes into a proton and emits a negative electron ($e^-$) and a neutrino ($\nu$).

pairs are emitted from the nuclei with considerable energy. In the $C^{11} \rightarrow B^{11}$ transformation, the electron-neutrino pair gets about one million electron volts; in the $C^{14} \rightarrow N^{14}$ case it gets 15,000 electron volts. These energies are set free because the end products, $B^{11}$ or $N^{14}$, have a lower energy than the initial nuclei, and the excess energy goes into the motion of the electron-neutrino pair. As we saw in the last chapter, a transition from a loosely bound system to a more tightly bound one always gives rise to a surplus energy. The nuclei of $B^{11}$ and $N^{14}$ have a more balanced neutron-proton ratio than the initial nuclei and therefore are more tightly bound.

There are many practical applications of radioactivity. For example, radioactive nuclei are of considerable importance in medicine because they produce the energetic electrons that affect living tissues. With modern accelerators it is relatively easy to produce radioactive nuclei. Nuclei with an abnormal surplus of protons or neutrons are obtained simply by bombarding ordinary nuclei with protons or neutrons. Some of the resulting radioactive isotopes have half lives of a few seconds, some a few hours or years; some even have half lives of billions of years. These long-lived ones need not be produced artificially; they are found on the surface of the earth, radium being a well-known example. They were produced at a time when the material of the earth was subjected to natural bombardment by protons or neutrons from some big star explosion. Because of their long half life, they are still with us.

Radioactivity[25] is a change from an unbalanced nucleus to a more stable one of different charge. The change is transacted by a neutron transforming into a proton, or vice versa. It is accompanied by an emission of an electron or positron together with a neutrino. This process involves a special kind of interaction among protons, neutrons, electrons, and neutrinos. It is called the *weak interaction*. It can be regarded as a fourth force of nature besides gravity, electromagnetism, and nuclear forces, because those three interactions cannot account for the transformations of radioactivity.

The weak interaction is indeed very weak. After all, we are dealing here with very slow processes. The years, hours, even seconds required for radioactive decay are very long periods of time from the point of view of a nuclear system, where motions are extremely rapid. These time intervals are very much longer than the natural time interval in nuclei, the "nuclear year"—approximately $10^{-23}$ seconds—in which a nu-

25. The term "radioactivity" includes another phenomenon that has nothing to do with the one we have described. Some heavy nuclei, such as uranium or thorium, are slightly unstable and, after very long time periods, decay by expelling an energetic alpha particle. This particle is identical with the helium nucleus. Rutherford used this kind of radioactivity for his experiments with alpha particle beams. Before the invention of accelerators, this was the only source available for high energy particle beams.

cleon moves around an orbit within a nucleus under the influence of nuclear forces. Rutherford once said radioactive transformations are so slow that they practically do not occur at all! But still they do exist. Even a solitary free neutron lives only ten minutes when it is not built into a nucleus. It transforms itself spontaneously into a more stable proton with simultaneous emission of an electron and a neutrino. As part of a nonradioactive nucleus, however, a neutron is as stable as a proton.

**Nuclear Energy, Nuclear Burning**
The heat of burning coal comes from the union of carbon and oxygen atoms, forming molecules with stronger bonds than the ones with which they were bound before. Energy is released whenever atoms join together and form a strongly bound unit. Can we apply the same principle to the bonds within the nucleus? Energy should be produced when neutrons and protons get together and form a nucleus. A nuclear fire should exist and be much more powerful than any ordinary fire, since the energies involved in nuclear phenomena are a hundred thousand times greater than the energies occurring in the electronic orbits of the atoms.

Let us consider a simple example of nuclear burning. The nucleus of helium consists of two protons and two neutrons, bound together by nuclear forces. The nucleus of carbon consists of six protons and six neutrons bound together tightly; hence we can think of carbon as being three helium nuclei in a close bond. If one could press three helium nuclei into close contact so that the nuclear forces began to act between them, they would snap together, forming a carbon nucleus and releasing large amounts of energy. Helium, therefore, should burn in a nuclear fire to carbon.

Why does ordinary helium here on Earth not burst into

nuclear flames? In ordinary circumstances it is extremely difficult to get three helium nuclei close together. First, they are surrounded by electrons. Second, being positively charged, they repel each other. Only at extremely high temperatures, billions of degrees, would the electrons be torn off and the nuclei have enough energy to overcome electric repulsion and collide with each other. Such are the temperatures required to ignite the helium fire, which, once kindled, would produce enormous amounts of energy and be a million times hotter than ordinary fire. We believe today that in the center of some stars such helium fires are burning and supplying the star with energy for its radiation. The upper left corner star of the constellation Orion is such a one.

There are other kinds of nuclear fire. An important one is the burning of heavy hydrogen. Heavy hydrogen is an isotope of ordinary hydrogen. Its nucleus, which is called *deuteron,* is composed of one proton and one neutron held together by the nuclear forces. Brought in close contact, two deuterons would snap together and form a tightly bound helium nucleus,[26] two protons plus two neutrons. Hence heavy hydrogen burns, and the ashes are helium. This nuclear fire also needs very high temperatures for ignition, but not as high as the helium fire. (The repulsion between deuterons is weaker than between helium nuclei.) Heavy hydrogen burning has been achieved by man, but so far only for destructive purposes in hydrogen bombs.

A most important nuclear fire is the burning of ordinary hydrogen. (See figure 52.) We believe that this is the kind of

26. Detailed studies have shown that the deuterons do not directly form a helium nucleus, as described in the text. They first collide with another proton to form a helium isotope $He^3$, which consists of two protons and one neutron; then two of these isotopes unite to form an ordinary helium nucleus $He^4$, freeing the two additional protons.

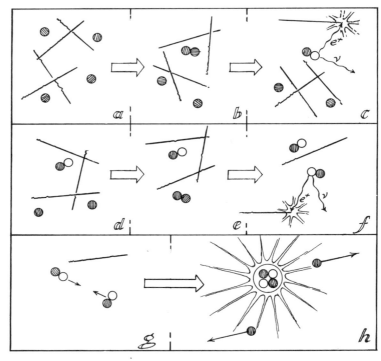

Figure 52
Eight stages of the burning of hydrogen to helium. (a) Four protons (hydrogen nuclei, drawn as small, filled circles) and four electrons (straight lines symbolizing their fast motion). (b) Two protons close, forming a diproton momentarily. (c) One of the protons changes into a neutron (open circle) with the emission of a positive electron and a neutrino. The positive electron hits a negative electron and annihilates in a light explosion (see section on subnuclear phenomena of this chapter). (d) A deuteron and two protons. (e) The second pair of protons forms a diproton. (f) The diproton changes over radioactively into a second deuteron as in (c). (g) The two deuterons hit. (h) They form a helium nucleus. The energy released is partly radiated (halo), partly transferred to other protons. The electrons are left out.

fire that burns in the interior of the sun and keeps the sun hot by providing the energy supply. It is not obvious how ordinary hydrogen can give rise to a nuclear fire, since its nuclei are pure protons, and neutrons are needed to form higher nuclei.

Here the mysterious process of radioactivity sets in. In very large amounts of hot hydrogen, it will happen from time to time that two nuclei—two protons, that is—come close together and temporarily form a nucleus consisting of two protons, a *diproton*. This entity is not very stable, but simple calculations show that once in a while one of the protons changes radioactively into a neutron, and a deuteron (proton plus neutron) is formed as the end product. The deuterons then burn to helium as previously described.

By this indirect process, ordinary hydrogen in large amounts and at high temperatures burns into helium. It is a slow process because deuterons must be made first before the actual burning occurs. The fuel supply for the fire trickles in bit by bit. As a result, which we can regard only as benign, the hydrogen fire in the sun generates and will continue to generate a steady heat supply over many billions of years, without any danger of sudden explosions.

The basic process of nuclear burning, the formation of larger nuclei from smaller ones—carbon from helium, helium from deuterons—is called *fusion*. There is also another process with which one can gain energy from nuclei, and this we call *fission*. The fission process is not very important in our understanding of the universe; it is not a large-scale source of energy in nature. Recently, however, man has managed to make use of it, both for energy production in nuclear reactors and for his own destruction in atomic bombs.

The nuclear forces, as we have seen, keep neutrons and protons together in a nucleus. But there is also another disruptive force acting—the electric repulsion between the protons. In all existing nuclei the binding effect of the nuclear forces is greater than the disruptive effect of the electric repulsion. If it were not so, nuclei would not exist. The nuclear forces, however, are very short-ranged; they act only if the particles are very close together, whereas the electric repulsive effects act over large distances. If one could split a nucleus in such a way that the halves were separated by an amount, which, though small, would be greater than the range of nuclear forces, the two parts would no longer stick together but fly apart, driven by the electric repulsion.

Normally it is very hard to "split" a nucleus, but some of the very heavy nuclei, such as a certain uranium isotope or the artificial element plutonium, are on the verge of falling apart. A hit by a single neutron suffices to split such an atom into halves that no longer stick together. The halves fly apart with considerable energy, heating the neighborhood to a high temperature. That is the process of fission. The process is so violent that usually one or two neutrons are chopped off when the nucleus splits. These chopped-off neutrons are of great importance. In a large chunk of fissionable material only one neutron is needed to start a reaction. The one neutron splits the first nucleus, the chopped-off neutrons go on producing further splittings and so on, until a large part of the material has split. This process we call a chain reaction. It can proceed only if the block of material it is contained in so large that, on the average, the chopped-off neutrons do not leave it before hitting another nucleus. The minimum size required is called the critical size; it is usually a few pounds of material. Any chunk of fissionable material larger than the critical size would develop a chain reaction when hit by one neutron, and thus would produce enormous amounts of energy. Atomic reactors are devices in which the amount of fissionable material is kept exactly at critical size; the energy and heat production can be regulated for practical uses.

The two halves of a nucleus produced in fission are themselves smaller nuclei, but their proton-neutron ratio is abnormal. In most cases they have too many neutrons, and therefore they are radioactive. This is why a nuclear reactor is such a prolific producer of radioactive material.

**The Quantum Ladder**
We can summarize what we have learned about the structure of matter by following the fate of a sample of atoms as we raise their temperature progressively from room temperature to that typical of the interior of stars. We choose as our sample a gas of sodium atoms. Most gases consist of molecules, but a few elements, such as neon, sodium, and lithium, do not readily form molecules in gaseous form. You are familiar with these atomic gases in their use as sources of light. The so-called neon tubes that bedizen our city streets are filled with atomic gases, neon or sodium or lithium vapor, each giving off a different color when an electric current is discharged through the tube. All these gases are composed of single free atoms.

Let us look at a tube containing sodium vapor. When the electricity is switched off, the temperature of the gas is the same as the temperature outside. At that temperature the heat energy, the energy with which the atoms move around, is about $1/40$ electron volt. This energy is far below the excitation threshold of sodium atoms; so when the atoms hit each other or the walls, they bounce off like hard billiard balls without changing their quantum state. At this temperature the atoms act like elementary particles: they do not show any internal structure. Their electronic pattern remains fixed and unchanged; all the atoms are exactly identical.

Let us raise the temperature of the gas by sending an electric discharge through the tube. When the energy transferred to the atoms by the electric discharge becomes higher than the threshold of excitation, other quantum states are excited. The atoms then emit their own characteristic light when they fall back to their lowest quantum state. Sodium atoms radiate yellow light; lithium atoms red light. The variety of city lights is based upon these typical colors. The excitation of atoms to higher quantum states is the beginning of a breakdown of atomic identity. No longer are all sodium atoms alike: some are in the ground state, some in other states.

We now raise the temperature further, so that the energy of the collisions between the atoms is very much higher than the excitation threshold—so high, in fact, that the electrons are torn off the atoms. Then the quantum states are all destroyed, and the electrons move like particles without any special wave patterns. We produce the plasma state of the gas, in which electrons and atomic nuclei move around in violently disordered motion. No two motions are exactly equal; the light emitted by the plasma has no characteristic frequency; it is unspecific heat radiation. However, the atomic nuclei and the electrons, though separated, still maintain their individual identity. They are the elementary particles of the plasma.

Now let us go to a still higher temperature, much too high for the laboratory, where ordinary containers such as glass or metal tubes would disintegrate. It is so high that the particle energies are beyond the stability limit of nuclei. Such temperatures exist only in the centers of stars, or, for a few milliseconds, in nuclear explosions. Now the nuclei lose their identity, some are excited to higher quantum states and emit their characteristic radiation—highly energetic gamma rays. Let us raise the temperature even further, to where the energy becomes so great that the nuclei fall apart into their constituents. All nuclear individuality is lost, and the material forms a disordered gas of protons and neutrons mixed with electrons that were torn off the atoms at much lower temperatures. Under these conditions matter is reduced to a mixture of three elementary particles: protons, neutrons, and electrons, without order.

The sequence we have outlined here we call the "quantum ladder." The sequence is established by a gradual increase of energy transfer. At our starting rung of that ladder, matter is composed of atoms as individual units whose inner structure is inert and rigid, moving about as billiard balls do. At the next rung the atoms are decomposed into electrons and atomic nuclei, and these particles are now individual units, inert and rigid. At the third rung the nuclei are decomposed into neutrons and protons; the units of matter in that stage are protons, neutrons, and electrons.

The existence of the quantum ladder has made it possible to discover, step by step, the structure of the natural world. When we investigate phenomena at atomic energies, we need not worry about the internal structure of the nuclei, since the nuclei cannot be excited with atomic energies. When we study the mechanics of gases at normal temperatures, we need not worry about the internal structure of the atoms, since atoms cannot be excited at room temperature. In the former case we can consider the nuclei as identical, unchangeable units—that is, as elementary particles. In the latter case each atom may be considered as an unchangeable unit. Thus the observed phenomena are simpler, and we can understand them without any knowledge of the internal structure of the constituents, which behave as inert units.

We can extend our quantum ladder to lower energies. When we cool down the sodium gas to low temperatures, the sodium atoms first form liquid sodium; at still lower temperatures they aggregate in a regular array, crystallizing and forming sodium metal. In other materials the step downward from the atom is more interesting. Isolated atoms exist in most materials only at temperatures we find in flames. At ordinary temperatures, most atoms (though not the atoms in sodium, lithium, or neon gas) join in groups and form molecules, which represent the next lower rung in our quantum ladder. Molecules represent individual specific entities whose stability threshold is lower than that of atoms because of their larger size. It is easier to decompose a molecule into its atoms than to tear the atom apart into its nucleus and electrons.

It is interesting to observe parallel phenomena at different steps of the quantum ladder. We find, for example, energy production when atoms join in molecules—the chemical fire—and energy production when smaller nuclei are fused to larger ones—the nuclear fire. These are two ways of burning, very dissimilar in the amount of energy involved but similar in principle; one occurs at the molecular level, the other at the nuclear level.

At the next step down the quantum ladder are the macromolecules; they are combinations of many ordinary molecules in special arrangements. Under certain conditions, macromolecules assemble in the form of large units that exhibit most astonishing properties, of which you will hear more in the next chapter. This is the rung of the quantum ladder where life occurs.

The last and lowest rung is established by matter at very low temperature. Almost all substances crystallize when cooled sufficiently; they form regular arrays of atoms or molecules. Heat motion disappears and total order is established—the order of complete immobility.

When we come to the lowest rungs of the quantum ladder, macromolecules and crystals, the size-stability relation must be applied with some caution. Since macromolecules and crystals are very large objects, one might infer that they would be extremely unstable. Indeed, they are easily altered in many ways. For example, macromolecules have no stiffness; they can be bent and folded with very little application of energy. Crystals can be brought into internal vibrations with extremely little energy; ordinary sound waves would do it. The important structural properties, however, such as the atomic structure of the macromolecules or the regular atomic arrangement of the crystal lattice, are very stable in spite of the large size. Just as the strength of a long flexible chain is the strength of the individual links, the strength of the macromolecule is that of the individual bonds that join one atom to the next one; that strength does not lessen as the molecule gets bigger.

Matter takes different forms according to the step on the quantum ladder. (See figure 53.) The lower the step, the higher the organization and differentiation of matter. Each step downward allows matter to settle in specific forms, which become more varied the farther down we go. At the highest rung, we have mentioned that protons, neutrons, and electrons move without any order. At the next lower step, the plasma, protons, and neutrons fall into the ordered pattern of atomic nuclei but electrons are still in disordered motion. Farther down, the electrons join the nuclei and form atoms; they fall into their typical atomic wave patterns.

At the next lower level atoms join in molecules. The differentiation becomes extensive; there are countless ways of combining atoms to molecules, each corresponding to another well-defined substance. The level of macromolecules contains even more variations—it is the step at which living matter occurs in its various forms and as organisms; it is the condition of matter with its widest possibilities. The energy ex-

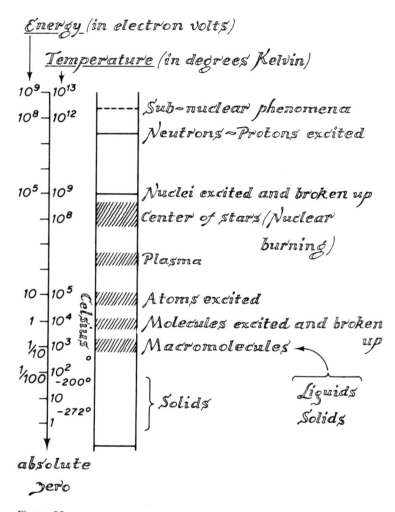

Figure 53
Quantum ladder. Phenomena are listed at appropriate energy levels. The temperature scale is measured in centigrades starting at zero, not the freezing point of water, but absolute zero.

changes at this level are low enough for the existence of large complexes of molecules, cells, and organisms, and high enough for the stimulation of growth and development of these entities. At the lowest rung, all variety, all differentiation is frozen into unchanging patterns of crystallized substances.

## Subnuclear Phenomena

Is there a step in our quantum ladder higher than the state of isolated protons and neutrons? The answer to this question brings us to the frontier of modern elementary particle physics.

Let us recall that energies of many millions of electron volts are needed to decompose the atomic nucleus. Modern high energy research has not stopped at this limit. In the last decade, particle accelerators have been built that attain many thousands of millions of electron volts. The unit of energy now used for those giant accelerators is no longer MeV (million electron volts) but GeV (giga-electron volts), which is a billion electron volts. At the Brookhaven National Laboratory on Long Island, New York, there is an accelerator in operation that produces proton beams up to 500 GeV. In Geneva, Switzerland, a united effort of most western European countries made it possible to construct several giant proton accelerators, one reaching 30 GeV, another 400 GeV. At the same laboratory there is also an interesting facility that makes two proton beams of about 30 GeV collide head on. Such a collision produces effects much more violent than an ordinary accelerator of two times 30 GeV would make when it hits a target. According to Einstein's mass-energy relation the mass of a rapidly moving proton is much higher than a proton at rest. Thus, in a colliding beam facility two "massive" protons hit each other, whereas in an ordinary accelerator only the accelerated proton is massive; the proton in the target has its usual mass.

In addition to accelerators that produce proton beams, a number of *electron* accelerators have been constructed, the largest of which is in Stanford, California, with a beam energy of 20 GeV. There are also colliding electron beam facilities at the Stanford laboratory and in Hamburg, Germany.

Study of the structure of the nucleus no longer is the principal use of these machines; rather, it is the study of the structure of protons and neutrons. Here a step is taken to the next higher rung of our quantum ladder. Have nucleons a limit of stability, too? Is there an energy above which some internal structure within the protons and neutrons themselves becomes observable? What are the proton and neutron made of?

One expects to find a very high stability limit for the proton and neutron because they are very small units. It is not surprising that they behave like inert and rigid particles at all energies except the very highest ones produced today. Not until several hundred million electron volts had been applied were indications of internal structure found.

When such tremendous energies impinge upon matter, one observes phenomena quite different from what is known in the atomic or nuclear realm. Indeed, completely new effects and new kinds of particles appear when matter is subjected to bombardments with high energy beams. A higher rung on our quantum ladder is reached. Much of what has been observed is not yet clearly understood, but we already begin to get a glimpse of what goes on at a deeper layer of the structure of matter. Following are four groups of phenomena that play an important role at these enormous energies.

1. The existence of antimatter
2. The excited states of the nucleons

3. Quarks as constituents of nucleons

4. The appearance of mesons

**Antimatter**    The discovery of antimatter is one of the most exciting episodes of modern physics. Physicists and chemists have wondered for a long time why positive and negative electricity plays such a different role in nature. When a current flows through a metallic wire, it is the negative electricity, the electrons, that are the moving parts. Positive electricity, does not participate in the flow of the current, since it is bound to the atomic nuclei. As Rutherford showed, positive and negative electricity in an atom are distributed in very different ways; the heavy nucleus at the center is positive, whereas light negative electrons surround the nucleus in fast motion. Why is nature so asymmetric in respect to the charges? Why are there no positive electrons and negative nuclei? So asked the physicists before 1927.

In that year the English physicist P. A. M. Dirac carefully studied the relations between the wave properties of matter and the requirements of Einstein's theory of relativity. By purely theoretical, logical arguments he concluded that the fundamental equations of quantum mechanics require the existence of an "antiparticle" to each particle. An antiparticle is very similar to its particle; it has the same mass, the same spin, but opposite charge and opposite magnetic properties. It was a great triumph for theoretical physics when only a few years later, in 1932, C. D. Anderson found the "antielectron"—the positron—among the particles that come to the earth's surface as cosmic rays.

The same holds for the proton or the neutron. The theory also required the existence of an antiproton, a particle much like the proton, but with negative charge. But if the neutron is uncharged, what would be the antiparticle to the neutron? The neutron carries a magnetic moment, the property that has the opposite sign in the case of the antineutron. It took much longer to discover the antiproton and the antineutron because of a circumstance that we now are going to discuss.

We come to the most fascinating point in regard to antiparticles. Dirac not only predicted the existence of antiparticles, he also predicted the occurrence of two most unusual processes: the annihilation and the creation of matter and antimatter. Indeed, these processes were observed later on. The particles and antiparticles are in a peculiar relation to each other. Whenever an antiparticle encounters a particle of the same kind—a positron hitting an electron, for example—a kind of explosion occurs and both particles disappear. They annihilate each other, and the energy contained in their masses and motions is transformed into some other form of energy, such as light or other kinds of radiation. Conversely, under certain circumstances, energy of some form can be transformed into a particle-antiparticle pair; for example, an electron-positron pair can be created by the action of light quanta. (See figure 54). These processes are the most impressive manifestations of Einstein's discovery that mass is a form of energy. According to his famous relation, $E = mc^2$, the mass of one electron corresponds to the energy of one-half million electron volts. The creation of an electron-positron pair, therefore, requires at least twice as much energy, which is more than 1 MeV. The mass of a proton or a neutron corresponds to about 940 MeV (almost 1 GeV); thus the creation of an particle-antiparticle pair requires about 2 GeV or more. This is why the antiproton and the antineutron were discovered as late as 1954 by Segre Chamberlain, Wiegand, and Ypsilantis at Berkeley, California; only then were accelerators of that power available.

From the evidence that antiprotons, antineutrons, and antielectrons exist, we expect that they combine to form antinu-

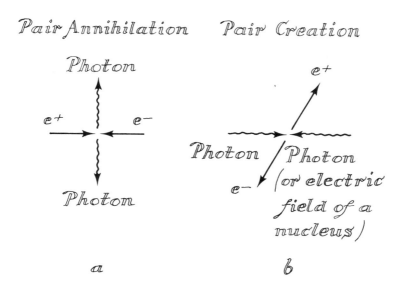

**Figure 54**
Pair annihilation and creation. In (a) an electron (e⁻) and a positron (e⁺) hit one another and are transformed into two photons. In (b) two photons hit one another and are transformed into an electron-positron pair. The second photon may be replaced by the strong electric field in the neighborhood of a nucleus. We then have pair creation by one photon close to an atomic nucleus. The photon provides the energy, it disappears, and the pair is created.

clei, antiatoms, antimolecules, and every form of antimatter. Indeed some simple antinuclei have been found as the end products of particle collisions. Obviously it will not be easy to assemble larger chunks of antimatter, since we live in a world of ordinary matter, and the slightest contact of an antiparticle with an ordinary particle of the same kind leads to annihilation of both.

A few words are in place about a very interesting little "atom" that can be made of a positron and an electron. These two particles attract each other because of their opposite electric charge. Therefore they form a structure analogous to the hydrogen atom (which consists of a proton and an electron), except that the proton is replaced by a positron. This structure was found by M. Deutsch at MIT and is called *positronium*. It is very similar to the hydrogen atom but has two important differences. First, it is very much lighter, since the positron has the same mass as the electron, whereas the proton is 1,836 times heavier. Second, it is a short-lived entity, since the positron and the electron eventually annihilate each other. This does not happen immediately; the time before annihilation is about a ten-millionth of a second. This is a very short time, but long enough to allow detailed observations of the properties of positronium. After all, the "atomic year"— the time an electron "runs around" the nucleus—is $10^{-16}$ seconds, which is smaller than the lifetime of positronium by a factor of a billion. Positronium is neither matter nor antimatter; it is a combination of both.

Antimatter plays an important role in the subnuclear realm. When particle beams with billions of electron volts impinge upon matter they are bound to create many particle-antiparticle pairs. Antimatter creation and subsequent annihilation are among the common events on this rung of the quantum ladder.

**Excited States of Nucleons**   The second group of phenomena appearing when matter is subject to extreme energies is one we might expect: excited states of the nucleons. When protons or neutrons are bombarded with particles of very high energies, higher than a few hundred million electron volts, they become "excited." The protons and neutrons, after collision, go into higher quantum states, just as atoms do when energies of a few electron volts are delivered to them, and just as nuclei do when energies of a few million electron volts are delivered to them. Let us go back to figure 50 where we saw two types of quantum states: atomic quantum states as exemplified by sodium atoms and nuclear quantum states as exemplified by sodium nuclei. We now face a third type of quantum state, that of the nucleon. Figure 55 shows a graph of some of the higher quantum states of the nucleon as they have been observed in the last twenty years. Note that the unit of energy in the nucleon spectrum is a billion electron volts. The energy distance between these new states is about a thousand times greater than the distances between the states of an atomic nucleus, which in turn are about a hundred thousand times larger than the atomic energy distances. The three different types of quantum states represent similar phenomena on three different steps of the quantum ladder.

The lowest state of the nucleon in figure 55 is the proton and the neutron. They appear to be on the same level. The two do not have exactly the same energy, but the energy difference between them is so small (1.2 MeV) that it does not show up on the scale of the figure.

The existence of numerous excited quantum states of the nucleon leads us to believe that there exists an internal structure of some kind. The same conclusion was drawn from the atomic and nuclear quantum states. The atomic quantum states point to the electrons as constituents of the atoms and reflect the different wave patterns of the electrons confined within the atom under the influence of the electric attraction. The quantum states of the nucleus point to the protons and neutrons as constituents and represent the different wave patterns of the nucleons within the nucleus under the influence of the nuclear force.

**Quarks** What do the quantum states of the nucleon tell us? Today we are beginning to decipher the secrets hidden

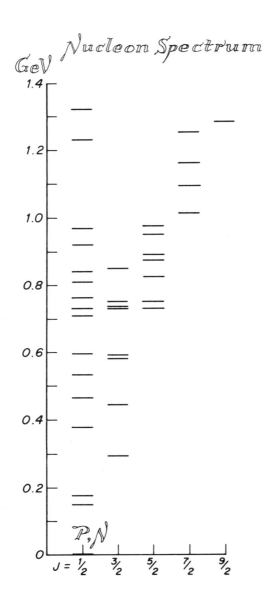

behind these states. It seems that they reflect the different patterns of a new type of subparticle that constitutes the proton and the neutron, the *quark*. Indeed, many properties of nucleons and their excited states can be understood by imagining that a nucleon consists of quarks. A group of physicists from MIT in Cambridge, Massachusetts, and the Stanford, California, electron accelerator laboratory (R. Taylor, H. Kendall, and J. Friedman) performed an experiment in 1969 that gave very convincing evidence for the existence of those subparticles. Their experiment was similar to Rutherford's experiment in which he found the nucleus of the atom. As we have described earlier, Rutherford directed a beam of $\alpha$-particles through a metal foil and found to his great surprise that a few of the $\alpha$-particles were strongly deviated from their paths. He explained this by the fact that they had a close encounter with a small massive electrically charged entity within the atom, namely the atomic nucleus. The MIT-Stanford group directed a highly energetic electron beam at a hydrogen target; some of the electrons hit and penetrated the protons in the hydrogen (remember, the proton is the nucleus of the hydrogen atom). They observed that a few of these electrons suffered strong deviations from their straight path. The quantitative evaluation of these deviations led to the conclusion that the electrons were deviated by some electrically charged entities within the proton that are much smaller in size than the proton itself.

---

Figure 55

The spectrum of the higher quantum states of the nucleon (proton and neutron). The lowest state (the lowest line in the $J = \frac{1}{2}$ column, referred to as $P$, $N$) is that of the proton and neutron. The states are grouped according to their angular momenta $J$, which, in units of Planck's constant $h$, are found to be $\frac{1}{2}$, $\frac{3}{2}$, $\frac{5}{2}$, $\frac{7}{2}$, $\frac{9}{2}$. There are also few states of even higher angular momenta that are not included here. The excitation energy is given in units of GeV, that is, billion-electron volts.

From all the evidence taken together, it seems most probable that the nucleon consists of three quarks. The number three was concluded from the character of the spectrum of excited states (figure 54), which, for the expert, shows many characteristic properties of a three-particle system.

There seems to exist several types of quarks, differing in electric charge, in mass, and in other properties. Some of these other properties are so novel and without parallel in the lower ranges of the quantum ladder that new names such as "strangeness" and "charm" had to be (somewhat whimsically) invented for these properties. The two most important quark types are rather prosaically called the u-quark and the d-quark. The electric charge of the u-quark is ⅔ of the unit charge $e$ ($e$ is the charge of the proton); the electric charge of the d-quark is $-⅓$ of $e$ (it is negative). The proton is thought to be the combination u, u, d, giving the total charge of $e$, as it should; the neutron is the combination d, d, u, which gives zero charge.

The quarks are supposed to be held together by a very strong force that physicists simply call the *strong interaction*. This force strongly holds together the three quarks within the nucleon. Indeed, it seems to be impossible to separate one of the quarks from the other two. This is a novel situation, since at the lower rungs of the quantum ladder it always was possible to separate the constituents and free them from the bond. The three quarks, however, seem to be permanently bound. Outside the nucleon, the effect of that force is much weaker. One might say that the force is almost completely contained within the nucleon. When two nucleons, each consisting of three strongly bound quarks, get near to each other, there is only a residual force acting between the nucleons, and that is the nuclear force that we discussed earlier; it is the force that keeps the nucleons together in the nucleus. The strong interactions keep the three quarks together within the nucleon so that no quark can be separated. A little of that force spills over and acts as an attraction between different nucleons, and that is the nuclear force.

This situation is not unlike that found with atoms. The electrons are tightly held together in the atom by the electric attraction of the nucleus. The atom itself, however, is electrically neutral; therefore there is not much of an electric force between atoms. Nevertheless, when atoms come near to each other, they influence one another enough that some of the electric forces spill over a bit and bind atoms together; thus, they form molecules or solid bodies. These "chemical forces" are weaker than the electric forces within the atom but strong enough to give rigidity to molecules and solids. In analogy, the nuclear force between protons and neutrons is weaker than the force that keeps the quarks together but strong enough to give rigidity to the nucleus.

**The Mesons**   We now come to another class of phenomena on this high rung of the quantum ladder, and that is the appearance of a new type of particles, the *mesons*. Whenever very energetic collisions of particles occur, one observes the emergence of mesons. Figure 56 shows a beam of protons with an energy of 25 GeV passing through liquid hydrogen. Two collisions are seen of a proton in the beam with a proton in one of the hydrogen atoms. We see a large number of mesons emerge. The mesons are rather massive particles, some of them charged, some of them neutral. They appear with different masses, the lightest about 300 times as heavy as an electron. There are heavier ones too; the heaviest discovered recently is almost five times as heavy as the proton.

Mesons are ephemeral entities. They don't live very long. After short time intervals, shorter than a billionth of a second, they transform themselves into other more conventional

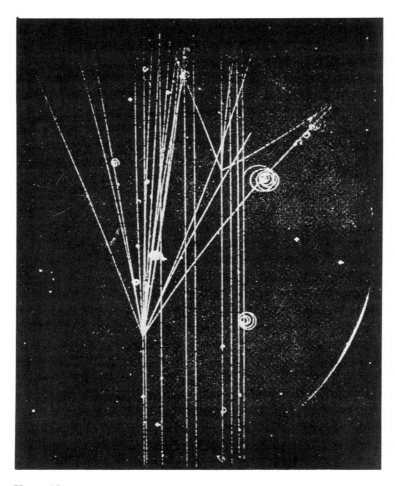

Figure 56
The creation of mesons by a beam of protons with an energy of 25 billion electron volts passing through liquid hydrogen. (Courtesy of CERN, Geneva, Switzerland.)

forms of particles. The charged ones become electron-neutrino pairs; the uncharged ones end up mostly as bunches of light quanta.

What are these mesons? In order to understand their structure, we have to invoke the fact that a quark, like any other particle, has an antiparticle, the *antiquark*. Today it is generally believed that a meson consists of a quark and an antiquark bound together by the strong interaction force. The different types of mesons are just different combinations of different types of quarks and antiquarks. As we have seen before, pairs of particles and their antiparticles are created when there is enough energy available. When energetic particles collide, the circumstances are favorable for the creation of quark-antiquark pairs—that is, of mesons. Again, the force between the members of the pair is such that it seems to be impossible to separate them. Quarks and antiquarks always appear as bound entities. The quark-antiquark structure of the mesons also explains why the mesons are ephemeral and have such a short lifetime. The quark and the antiquark annihilate each other and thus transform the meson into other forms of energy. A meson is neither matter nor antimatter; it is the quark-analog to positronium, the entity consisting of an electron and a positron, which we described earlier. The differences between a meson and positronium are the nature of the particles and the nature of the force that keeps them together before annihilation. In positronium the particles are electrons and their "antis"; mesons consist of quarks and their "antis." The force is electric attraction for positronium and the much more powerful strong interaction for mesons.

The production of mesons is a characteristic feature of the subnuclear realm. So much energy is exchanged between the particles on this rung of the quantum ladder that quark-antiquark pairs are profusely created and emerge as mesons.

Proton-antiproton pairs are also created, as well as other types of particle-antiparticle pairs. The subnuclear realm abounds with matter and antimatter created and annihilated in the highly energetic collisions that are needed to realize this realm.

The creation of so many entities poses a difficult problem for researchers in the field. Suppose you want to know what certain particles are made of. One of the methods to find out, which is used constantly by physicists, is to let the particles collide. Here is an analogy. Suppose you want to know what a Swiss watch is made of. A good method to find out would be to take two of them and hurl one against the other with enough energy to overcome their rigidity. Then you observe the parts emerging from the collision and you identify them with the constituents of the watch. This method is quite useful for atoms and nuclei. If the energy of the collisions is just greater than the breaking point, the emerging particles will be electrons and nuclei in the case of atoms and protons and neutrons, or combinations of them, in the case of nuclei.

But in the subnuclear realm the collisions must be more energetic in order to surpass the limit of stability of the nucleons. Under these conditions, many new particles and antiparticles are created. What emerges from the collisions is not only what was inside but what was created during the collisions. This circumstance makes the analysis of the internal structure more complicated. Of course, even in the case of the Swiss watches, not everything that emerges need be contained in the watches before the collision. What about the photons emitted by the sparks that may be produced when a collision is violent?

Having reached a new rung of the quantum ladder, anybody who thinks about the situation will ask the question: Are there further rungs? At each rung of the ladder, the next higher rung is quiescent. In the atomic realm, the nucleus remains unchanged and stable, since the energy exchanges are not sufficient to excite the nuclear quantum states. The internal nuclear dynamics lie dormant under these conditions. These are the conditions of matter on the surface of the earth in the environment in which we live. However, under much more violent conditions, such as the ones prevalent in the hot center of the stars, the nuclear dynamics becomes active and nuclear processes occur. The nuclear physicists were able to create such conditions in the laboratory in targets hit by particle beams that were accelerated by cyclotrons or other similar machines. They also created such conditions in nuclear reactors and in nuclear explosions. Still, the neutrons and the protons were left unchanged and stable under these conditions. The internal quark dynamics remained dormant. Only when particle beams with energies of gigavolts were applied was the limit of stability of the nucleons reached and the internal dynamics of the nucleons made to act up; excited nucleons were observed and mesons created. (See figure 57).

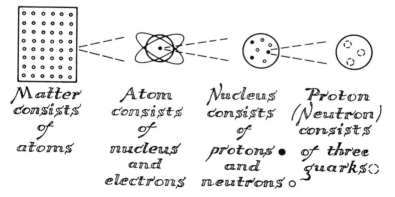

Figure 57
The different layers of the structure of matter.

Will there be further rungs of the quantum ladder? Is there an internal dynamics inside the quarks, or inside the electron, which is dormant in the presently available energy regions but sensitive to even higher energies? Are the several types of quarks that have been observed perhaps an indication of internal states of excitation in the quark? Are the so-called "heavy electrons," which have appeared in certain subnuclear processes, an indication of an internal structure of the electron? Today, nobody can answer these questions. Only further research in this fascinating frontier of science will show whether there is an unending series of rungs in the quantum ladder, each opening up new realms of material behavior, or whether there is a definite innermost structure of matter from which all other structures can be derived.

# 8 Life

In the preceding chapters we strove for an insight into the structure of matter. We examined atoms and nuclei and the various combinations of atoms into molecules. Conditions on the earth are such that most atoms are found in their specific lowest quantum state and aggregate to form molecules. The result is that we find on Earth so many materials with well-defined properties: minerals, metals, water, air, etc. But this does not hold on the surface of the sun. There the temperature is so high that molecules cannot exist. They would be torn apart immediately into atoms. Consequently we expect only elements and no molecular compounds on the sun, and everything in the form of hot vapors. We on Earth enjoy a much more varied environment, since we live in the midst of so many different substances in solid, liquid, and gaseous states.

Materials and chemical substances are inert and passive. They change their forms or chemical constitutions only under the influence of external causes. Air is moved by the heating of the sun, water by wind or by gravity, solids by mechanical or chemical influences such as wind and weather. Chemical processes are initiated by the heating and cooling caused by sun and weather and by the heat flow from the interior of the earth. A boulder field traversed by a creek may serve as an example. The sizes of the rocks in the creek bed vary from small grains to big boulders; they were shaped in the events that occurred when debris came tumbling down from some crumbling mountain. Each boulder consists of small crystals, whose structure and hardness are determined by the characteristic properties of silicon oxide, the substance of most of the rocks. The surface of the rocks shows traces of chemical reactions with oxygen in the air or with water from the creek or from rain. But for all the evidence of change, of upheaval or tearing down, of never-ending chemical activity, the stillness most impresses us. Nothing is moving except the water, which was lifted into the atmosphere by evaporation and now is being drawn into the valley below by gravity. A gust of wind may shift a few grains of sand or tumble a pebble, but these are mechanical movements that do not seem to affect the inert nature of matter.

There is something else on Earth, however, that does cause motion and change, and it represents a wholly new form in which matter appears. Wherever we look we find manifestations of *life*. The phenomena of life do not seem to fit at all into the framework of the events which we so far have come to expect from matter composed of atoms and molecules. Living matter is not passive and inert. It grows, it multiplies, it moves around on the ground, in the sea, and in the air; its activities seem to be determined by internal and not external causes. Living objects exhibit characteristic shapes and sizes contrasting sharply with anything made of ordinary matter. Of course the latter also exhibits typical and characteristic features, but the size and the shape of a boulder or a piece of metal depends on the special conditions of the immediate environment. Furthermore, when such an object is broken apart, its fragments still will be boulders. In living matter, on the other hand, the shapes and sizes recur and repeat and are largely independent of the incidental conditions of their environment. There exists an obvious unit of living matter—the individual organism. If it is broken apart its fragments are in most cases no longer such organisms. It makes definite sense to speak of 1,000 bacteria, 1,000 rosebushes, or 1,000 lions; these units are much larger than the natural units of matter, the molecules.

Chemical analysis has shown beyond a shadow of a doubt that living objects consist of the same kinds of atoms as nonliving things. In fact, living matter consists mainly of the four elements carbon, oxygen, hydrogen, and nitrogen, with traces of other elements, such as iron, phosphorus, and calcium. There is not the slightest indication that living matter contains any special material or that the laws of interaction between the atoms are different. The phenomena of life, therefore, must be the result of ordinary interactions between atoms and molecules, giving rise to very special molecules and combinations of molecules, and forming complicated structures that distinguish them strikingly from the molecules of lifeless matter.

Today we do not yet completely understand how the interaction of these molecules can give rise to the phenomena of life. In the last three decades, however, biological research has provided so many new insights into the molecular structure of life that we already have a vague idea of what goes on in living matter. The recent progress of our understanding of life is one of the great scientific achievements, comparable to Newton's and Maxwell's work and to the insights that quantum mechanics has provided. We have a special stake in the understanding of these structures, since not only are our own bodies living matter, but life in some form composes the most important part of our environment.

**The Molecules of Life**

Life exists in many shapes. Let us first look at a simple form of life, a bacterium.[27] (See figure 58.) It is a very small thing, only one ten-thousandth of an inch long, shaped like a sausage, with a skin and a jellylike substance inside. Such a unit is called a *cell*. In order to understand the essential features of this living object, let us compare it with a similarly shaped

27. We are describing here the bacterium *Escherichia coli*. (See figure 58). There are many different kinds of bacteria, and not all have the same properties.

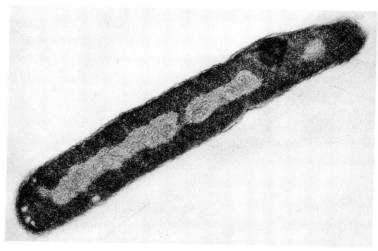

Figure 58
An electron microscope picture of a bacterium (*E. Coli*). Magnification 48,000 times. (From E. Kellenberger, A. Ryter, and J. Sèchaud, *Journal of Biophysical and Biochemical Cytology* 4 (1958).)

nonliving object, say a plastic sausage-shaped skin filled with a jellylike substance such as fat or gelatin. The walls and the inside of the nonliving object would be homogeneous; the inside and the container would consist of a large number of identical molecules of the same kind. The molecules of the plastic would make up the skin; the fat or gelatin molecules would make up the content. In a cell, however, the situation is vastly more complicated and vastly more differentiated. The units of material that the cell is made of are complicated combinations of large numbers of molecules, so-called *macromolecules*. A macromolecule is a combination of many smaller molecules of similar types that are joined together to form a chainlike structure. Later we will describe some of those macromolecules in greater detail. In a single cell there are very many, as many as five or six thousand, different kinds of macromolecules, each having a well-defined special structure.

But this complication is not the main difference. Let us put the plastic bag filled with fat or gelatin and the actual bacterium in a so-called nutrient solution—that is, a solution of mostly sugar, phosphates, ammonia, and traces of other chemicals. The plastic bag won't change much. Some of its contents might seep out through the pores of the skin and some of the solution might seep in. But the bacterial cell will change a lot. It will grow; more macromolecules will be formed inside the cell. The molecules of the solution seep through the skin into the cell where they are decomposed, and the atoms rearrange to form new macromolecules. When this process has gone on for a time, an even stranger event occurs. The cell divides into two, and each part starts growing for itself, also dividing into two cells when it is grown up. At the end, when all the nutrient material is used up, there will be many cells in the solution, all containing the same macromolecules that were in the original cell. The relatively simple molecules of the nutrient—sugar, phosphate, ammonia—are all transformed into the complicated macromolecules of the cells.

This is the process of life. The ability of replication is typical for life and does not occur elsewhere in nature. A living unit is capable of constructing another sample of itself with all its complicated details when the very much simpler constituents are available. Sure enough, a lifeless salt crystal also is able to grow and multiply in a saturated solution of salt; ice crystals of various shapes are formed when humid air is cooled down. But these crystals are combinations of simple molecules, and they are not exact replicas of an original crystal, either. All this is a far cry from the ability to combine simple molecules such as sugar, ammonia, and phosphate, and form exact replicas of the complicated molecular combinations of these simple constituents and, furthermore, put them together in the same way as in the original cell so that they again form a cell capable of growth and renewed replication. These are the wonders of life that we try to understand. Can they be explained on the basis of the ordinary laws that power the behavior of atoms and molecules in the lifeless world? Yes; most scientists believe so, and we will try to describe how it all comes about.

Living matter consists of complicated combinations of molecules, the so-called macromolecules. There are essentially two types of them in the cell, the *proteins* and the *nucleic acids.* The bulk of the cell is made up of proteins; nucleic acids are in the minority, but they play a decisive role.

Let us start with a description of the proteins. They are large units built up of amino acid molecules, the type described in chapter 6. The amino acid molecules are arranged like beads on a string, one following the other in a linear array, often as many as 1,000 in a row. A typical property of the macromolecules of life is: They are chains of smaller units, put in a certain well-defined order, that form long chains, in which one molecule follows the other.

The order in which these units are arrayed is of great importance. We find twenty different kinds of amino acids in proteins, some of which were described earlier and appear in figure 40. They bear such names as glycine or alanine, but we shall call them simply by the letters of the alphabet, a, b, c—and we need twenty letters for this. A protein, then, is described when we enumerate its amino acids in the order in which they are arranged. (See figure 59.) Any array of letters like c, f, m, u, a, d, etc., would identify a particular protein.

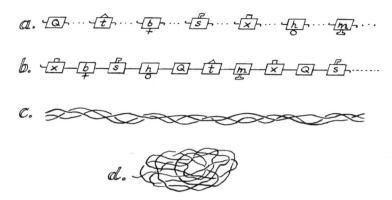

Figure 59
Protein structure. (a) Different amino acids. The two hooks symbolize the carboxyl group on one side, the amino group on the other. They hook up and bind the amino acids together. (b) An amino acid chain. In rows (c) and (d) this chain is represented as a line. (c) A fibrous protein. The chains wind like the strands in a rope. (d) A globular protein. The chains are coiled up in a skein.

Then thousands of letters would be needed in order to describe the big ones among them. There are uncounted ways of arranging twenty different kinds of amino acids in a row of 1,000. Each arrangement is another protein. We can get a feeling for the immensity of the number of the possible proteins by observing that 1,000 letters take up approximately two thirds of a page of a book. Each way of filling these pages with letters, whether the sequence contains actual words or not, corresponds to another kind of protein.

The proteins found in living matter are only a small part of all possible proteins. They comprise only the "sensible" combinations of amino acids, the combinations used in the structure and chemistry of cells. They correspond, as it were, to the book pages containing meaningful sentences. Still, the number of such sensible proteins is immense. For example, the proteins that make up the human skin are slightly different in every human being. This is why it is impossible to graft human skin of one person on another except when the two are identical twins.

The bacterial cell is one of the simplest living units and therefore contains a much smaller variety of proteins—"only" about 5,000 different kinds. The proteins differ from each other in many ways. Some are stiff and look like fibers; they serve as material for the skin of the cell, for internal divisions and membranes (not unlike the proteins in the human skin). Other proteins are pliable, so much so that the long chain of amino acids is all coiled up in a skein. Called globular proteins, they are able to move around, and they make up most of the jellylike content of the cell. These latter proteins are chemical agents; they can engage in chemical reactions, which are needed in the process of growth and in the digestion of food. One needs quite complicated mechanisms for such specialized tasks; that is why some proteins are very intricate combinations of molecules.

Next we come to the second type of macromolecules, the nucleic acids. They represent only a small part of the cell but, as we shall see, the decisive part. The most important nucleic acid is deoxyribonucleic acid—DNA for short. (See figure 60.) DNA is again a chain of units arranged in a linear array, one after the other. The units are not amino acids but molecules called nucleotides. There are only four different kinds: cytosine, guanine, thymine, adenine. We are not interested here in the details of their structure;[28] they contain atoms of carbon, nitrogen, hydrogen, oxygen, and phosphorous. Let us call them simply C, G, T, and A. The units of the chain forming our macromolecule DNA are actually a little more

28. Actually the nucleic acid chain is not merely a sequence of nucleotides. The outside frames of the spiral (the double lines in figure 61) are made of phosphate molecules and a certain type of sugar.

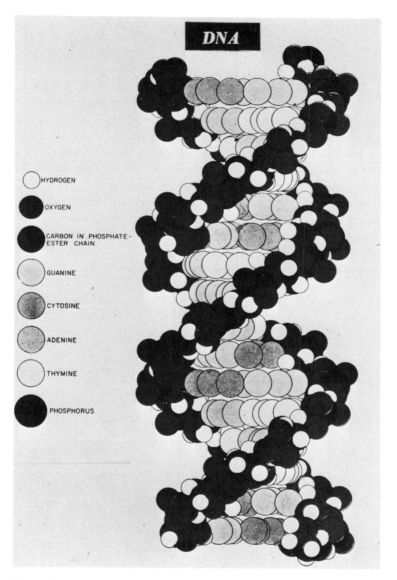

**DNA**

HYDROGEN

OXYGEN

CARBON IN PHOSPHATE-ESTER CHAIN

GUANINE

CYTOSINE

ADENINE

THYMINE

PHOSPHORUS

Figure 60
This model of a portion of the DNA molecule shows by means of different shades and sizes the variety of atoms present and the complex spiral structure. (From A. Rich. *Reviews of Modern Physics* 31 (1959).)

complicated. They are pairs of nucleotides. The following pairs are used as beads in this chain: C with G and A with T.

Because of the arrangement of the pairs of nucleotides, it is perhaps better to describe the nucleic acids as a ladder rather than as a chain. (See figure 61.) Each rung of the ladder is one of the pairs. It makes a difference which of the nucleotides in the pair is on the right side and which is on the left side of the rung. Hence there are four different kinds of rungs: CG, GC, TA, and AT. They follow each other step after step in a well-defined order that characterizes the DNA-molecule ladder. In addition, this ladder is twisted in a spiral, so that the whole macromolecule looks more like a spiral staircase, each step being one of the nucleotide pairs. In living cells these

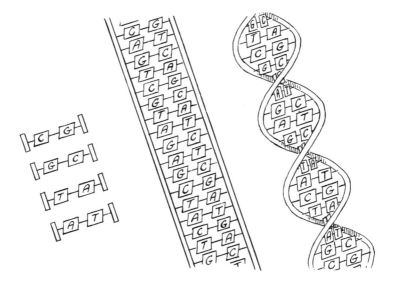

Figure 61
Schematic pictures of the DNA molecule. (a) The four units of the chains. (b) The ladder without twist. (c) The actual form of the twisted ladder.

molecules have enormous lengths—they contain as many as several hundred million pairs of nucleotides in a row. When found in the cell, the spiral is all coiled up in a tight skein. With the skein unfolded, the total length of the spiral ladder would be about an eighth of an inch in bacteria cells, and as long as several feet in human cells. (See figure 62.)

Here we must pause and think. What we have is a molecular structure of the size of inches or feet—a macroscopic size, as large as objects on our table. But still it is one single molecule. It is, of course, the large number of nucleotides that causes these enormous sizes; each pair of nucleotides is very small, as small as we expect ordinary nonliving molecules to be, say some $10^{-7}$ centimeters. But many hundred million in a row amount to lengths in meters.

There is some reason that the maintenance of life should need such long molecules. We shall see this soon in greater detail. At present let us be content to emphasize the tremendous variety of possible DNA arrangements.

We already have observed the enormous number of ways one can build up a protein chain of 1,000 amino acid beads when there are twenty varieties of beads. In the case of DNA, we have only four types of beads, but 10 to 100 million of them in a DNA-molecule ladder! It is important to realize that the restriction of bead types to four (instead of the twenty in the proteins) reduces the number of different arrangements, but not by very much. This reduction is vastly overcompensated for by the very much larger number of beads. But it is possible to have a written language with even two letters; this is done in the Morse code, where only dots and dashes are used. True enough, one needs on the average three or four Morse signals per letter, so that 1,000 signals would correspond to only a fifth of a page. But a DNA molecule (apart from having four and not two types of symbols) contains $10^7$

Figure 62
The DNA molecule as seen in the electron microscope. Microscopic plus photographic enlargement is about 20,000 times. (From A. Kleinschmidt, *11th International Congress of Surface Activity,* Cologne, 1960, vol. II.)

to $10^8$ rungs (many thousand times more than a protein), which would correspond to a book of 1,000 to 10,000 pages. Thus the number of different ways to build a DNA molecule is as large as the number of possible arrangements of letters (sensible and not sensible) in a book of 10,000 pages or more.

We soon shall see that this variety is connected with the variety of life; the arrangement of the four types of pairs in the DNA molecule is the book that tells the cell what to do and how to develop. An important question remains, however: How does one read this book?

## The Chemical Process of Life

Let us now come back to what we called the process of life, the growth of the bacteria cell and its division into two new ones when it is immersed in the nutrient solution containing mostly sugar, phosphate, and ammonia. It is a most interesting and puzzling process.

There must be in the cell of the bacterium a mechanism able to do two things: first, to build the twenty kinds of amino acid molecules and the four nucleotides from sugar, ammonia and phosphates; and second (this second step is much harder), to combine the amino acids in the correct order to form the thousand different proteins and to provide replicas of the nucleic acids in the process of division.

The first task, the fabrication of the "beads," is done, as we have mentioned before, by certain proteins in the cell. These proteins have the ability to decompose nutrient molecules after they have seeped in through the skin and rearrange their atoms into amino acids or nucleotides.

The second task, the arrangement of the beads in the correct order to form the new proteins or the new nucleic acids, is the one in which the large nucleic acid chains are involved. It is a very complicated process, and that is why it needs so many and such complicated proteins and nucleic acids. Only in the last two decades have the basic principles of this mechanism been discovered.

Let us start with the fabrication of the beads making up the protein chains, the amino acids, and the nucleotides. The raw material—the atoms that the beads are made of—are all found within the sugar, ammonia, and phosphate molecules, and in other salts contained in the nutrient solution. These molecules enter the cell through the pores of the skin, so all that is necessary is to decompose these simple molecules of the nutrient solution and put the parts together in the form of an amino acid or a nucleotide. It is the proteins that perform this essential step. A group of specific proteins is assigned to each of these tasks, such as the production of the different nucleotides and the different types of amino acids.

How do these proteins perform this task? They have the property of attracting the right molecules of the nutrient solution and of reshuffling the atoms such that they form the special amino acid or the special nucleotide. When the protein encounters sugar, ammonia, and phosphates, they become attached to the protein at certain specific places. These places are so arranged that the atoms within those materials, once attached, are forced to fall into the scheme of the amino acid or of the nucleotide to be formed.

All these processes are examples of the astonishing properties of proteins; they can perform and direct chemical reactions. Proteins with this ability usually are called *enzymes*. They are much more complicated than the simpler proteins that are not enzymes, which serve only to give the cell structure and regulate the flow of materials. Enzymes are able to build all the constituents of which the macromolecules are made by rearranging the atoms, or groups of atoms, contained in the nutrient. They can produce the beads, but they cannot put them together to form macromolecules. Later we will come to those most surprising mechanisms by which these beads are put together.

The enzymes can perform the task of making amino acids and nucleotides, but they need energy—energy to decompose the nutrient molecules, move the parts to the right places, and put them together with the correct bonds.

Where does this energy come from and how is it delivered to the enzymes? The sugar molecules of the nutrient solution in which the bacterium was immersed contain energy. We know that the burning of sugar can release much energy in the form of heat, when the sugar is transformed into carbon dioxide and water. The heat energy would be quite useless for the purpose here, however, since it is irregular random motion that cannot be used for purposeful molecule construction.

We mentioned in the sixth chapter that it is possible by special "rigging" to transform the energy of burning into energetic quantum states of certain molecules instead of letting it turn into heat. Here in the cell this rigging is done by specific proteins. How do they do it? These proteins are able to attract a sugar molecule to their surface. There the molecule is forced to fall apart into groups of atoms, which are rearranged by the protein to form carbon dioxide and water. Remember that this process is equivalent to "burning." What happens with the energy released in this process? The protein attracts molecules of another kind, which are kept close to the decaying sugar. These molecules (always present in the cell) are called adenosine triphosphates, ATP for short. They can assume two quantum states, one of lower and one of higher energy, and therefore can act as storage for the energy extracted from the sugar. Whenever sugar is decomposed the protein transfers the released energy to the ATP molecules.

The ATP molecules are lifted into their higher quantum states. If energy is needed somewhere else in the cell for a molecule synthesis, the ATP molecule will get there and deliver the energy by falling back into its less energetic quantum state.[29]

The ATP energy carriers have another advantage—they carry very small amounts of energy. The energy liberated by the decomposition of one sugar molecule is divided up among about forty ATP molecules. The energy is changed into small coins, so to speak, and can be distributed more readily to the many activities for which it is needed.

Up to now we have seen how the beads of macromolecules, the amino acids, and the nucleotides are produced. If we compare the beads again with letters, we can say the enzymes produce the letters; but they cannot put them together. The letters are there, but where is the author to compose the words and sentences?

## The Master Plan of Life

The step of putting amino acids together to form proteins contains all the secrets of the life of a bacterium, since, as we have seen, it is the different types of proteins that perform all the important steps in the chemical life of the cell. Where in the cell is the master plan hidden for each of the many thousand proteins, the plan that determines the order in which the amino acids follow each other along the string? Let us recall that each one of the proteins is a string of roughly a thousand amino acids (sometimes more, sometimes less) and that if we ascribe to each type of amino acid one letter of the

29. In fact, the state of lower energy is the molecule adenosine diphosphate (ADP). The loading with energy is accompanied by acquisition of another phosphor atom; the state of higher energy is then adenosine triphosphate (ATP).

alphabet, the array of the amino acids corresponds to an array of a thousand letters, as many as we find on about a page of this book. To specify the order of amino acids in 5,000 proteins we would need several thousand pages of a book like this one. Where is the information to be found in the cell? We only need to remember that the nucleic acid macromolecules have in them the possibility of expressing the content of many thousand pages of a book. The order in which the four types of nucleotide pairs are arranged along the winding spiral could give us that information. There are just enough possibilities in the order of the steps of the nucleic acid in the bacterium to determine the 5,000 proteins that make up the bacterium.

The big question poses itself: How does the order of nucleotides in the nucleic acid determine the order of amino acids in the proteins? How is the information contained in the steps of the spiral transmitted to the newly formed proteins? How can the cell "read" this book of many thousands of pages and follow the instructions when it grows and divides?

Today, we don't know yet in detail how such transmission of information occurs. But we understand the essential steps of this process. Three subsequent nucleotides always characterize an amino acid. Biologists have tabulated this code in the form of a small dictionary. We remember the four different kinds of nucleotides: C, G, T, and A. To each amino acid there corresponds a triplet of nucleotides. For instance, GGC means the amino acid glycine, CGG means arginine, CAA means glutamine, etc. In most cases one amino acid is expressed by more than one triplet. For example, not only CAA, but also GAG means glutamine. Glycine is expressed by four triplets, GGG, GGC, GGA, and GGT. But for any triplet there is one and only one amino acid. There are three exceptions: The triplets TGA, TRA, TAG do not refer to any amino acid.

They probably are used as symbols for punctuations; they may mean "start here, a protein begins" or "stop there, the protein is finished." The following table shows the twenty amino acids with the code triplets. It represents the "dictionary" for the cell.

| Code Triplets | |
|---|---|
| Glycine | GGG GGC GGA GGT |
| Alanine | FCG GCC GCA GCT |
| Glutamic Acid | GAG GAA |
| Aspartic Acid | GAC GAT |
| Valine | GTG GTC GTA GTT |
| Arginine | CGG CGC CGA CGT AGA AGG |
| Proline | CCG CCC CCA CCT |
| Glutamine | CAG CAA |
| Histidine | CAC CAT |
| Leucine | CTG CTC CTA CTT TTA TTG |
| Serine | TCG TCC TCA TCT AGC AGT |
| Threonine | ACG ACC ACA ACT |
| Lysine | AAG AAA |
| Methione | ATG |
| Isolencine | ATC ATA ATT |
| Tryptophane | TGG |
| Cysteine | TGC TGT |
| Tyrosine | TAC TAT |
| Phenylalamine | TTT TTC |
| Asparagine | AAC AAT |
| Punctuation (Stop) | TGA TAA TAG |

The nucleotides in the nucleic acid are aligned in such a way that the sequence of triplets corresponds exactly to the sequence of the corresponding amino acids in the proteins. If, in a protein, for example, the sequence glycine-arginine-tryptophane appears, we will find the sequence GGG-CGG-

TGG in the nucleic acid. Thus, by following one side of the spiral ladder of the DNA and paying attention to the sequence of nucleotides, we will find the description of all proteins of the cell, one after another. The description of a protein with 100 amino acids requires 300 subsequent nucleotides. The millions of nucleotides in the DNA molecule are therefore sufficient to describe all proteins the cell needs.

But how is this "description" used? How does nature read the nucleic acid book and turn this information into a real protein? In a very simplified way, we can picture protein formation as follows. First, copies are made of those parts of the DNA that contain the information regarding one specific protein. These copies are in the form of another nucleic acid, RNA (ribonucleic acid), which consists of a chain of single nucleotides, not pairs, as in DNA. Each of these copies is, of course, much shorter than the original DNA, since it contains only the part concerning one protein. We call them *messenger RNA*. There is a different one for each protein to be formed. After the RNA copies are made from the DNA, they move away from the cell nucleus, the part of the cell where the DNA is located, and go to the so-called *ribosomes*, special parts of the cell devoted to protein production. The different amino acids are all available there and ready to be assembled. The ribosome is an interesting little piece of machinery. The messenger RNA is "inserted" into it, and, when it slides through, the ribosome "reads" the code, collects the corresponding amino acids, and attaches them in the order they are read off the messenger RNA. For reading and assembling, each amino acid floating around near the ribosome has a "tag" attached to it, naming the three nucleotides that fit snuggly to the triplet composing the amino acid. For example, the triplet CGA means arginine; therefore arginine has a tag made of GCT, since G fits snuggly with C, C with G, and T with

A. (Remember, the snuggly fitting pairs were the rungs of the ladder in DNA, as described earlier.) Now if a given triplet of messenger RNA slips into the ribosome (say CGA), the amino acid (arginine) with the corresponding tag (GCT) will be caught because of the snug fit. All the ribosome has to do is remove the tag and attach the amino acid to the protein under production. (See figure 63.)

Let us summarize. The cell is made of different kinds of proteins; the simple ones make up the skin and the structural framework. Others burn sugar and produce the energy-bearing ATP molecules. The most complicated proteins produce amino acids from the chemicals provided by the nutrient. The cell also contains the nucleic acid macromolecules that contain the instructions of how to put the amino acids together so that they form proteins. Then there is a mechanism—the ribosome—that receives these instructions from the nucleic acids and constructs the proteins.

When the cell reaches a certain size, some largely unknown factors cause a rearrangement of the proteins, and the cell divides into two equal and smaller cells. At that stage it is necessary to duplicate the all-important nucleic acids, since each cell needs a set of them for further growth. The duplication of such a long and well-ordered molecule is not an easy process. We don't know exactly how nature does it, but we have a rough idea of what is going on. Here is a simplified version of the mechanism that duplicates the spiral ladder making up the long DNA molecule.

Each rung of the ladder, you will remember, is a definite pair of nucleotides. At cell division time the ladder is cut in two parts lengthwise by breaking each rung in the middle. (See figure 64.) The two partners of the pairs constituting the rungs simply separate, and two half ladders are formed. Now we must remember that the cell at that point contains free-

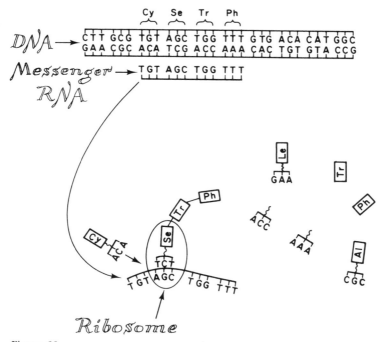

Figure 63

Mechanism for protein production. On the top of the figure we see a DNA molecule. For clarity's sake, the spiral winding is omitted and it is shortened so that it contains only ten triplets of nucleotides on each strand, the lower being paired with the upper. Directly below, we see a copy of part of the upper chain of the DNA. This is the messenger RNA. It contains the information for one protein. In this example the protein has only four amino acids: phenylalamine (Ph), tryptophane (Tr), serine (Se), cysteine (Cy). The codes are TGT for Cy, AGC for Se, TGG for Tr, and TTT for Ph.

In the lower right of the figure we see single amino acids. Some of them are attached to a "tag" containing the triplet that pairs with their code. The messenger RNA slides through the ribosome. The code triplet AGC is in place and attracts the serine because it is attached to TCT, which pairs with AGC. The ribosome then attaches serine to the protein that already contains Ph and Tr. The Cy attached to ACA waits for the next step.

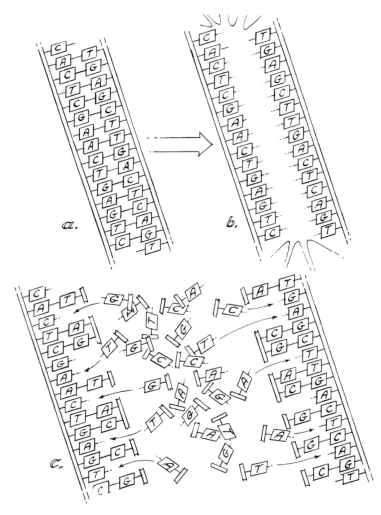

Figure 64
Multiplication of DNA. (a) DNA molecule without spiral winding. (b) The lengthwise split of DNA. (c) Each half of a DNA collects the correct nucleotides and forms a new complete DNA molecule, identical to the original one.

swimming nucleotides, which were produced by special proteins (enzymes). Each half rung then finds the corresponding nucleotide in the cell and again forms a full rung. At the end of this process two complete ladders with exactly the same order of steps are established; after division each goes into one of the newly formed cells.

We now have a vague idea how a bacterium grows and multiplies. When a few bacteria are put into a nutrient solution, the process of life transforms the simple but energy-rich nutrient molecules into complicated molecules of more and more bacteria until the nutrient is used up. This process is possible only because of the existence of the DNA macromolecules, which not only determine the structure of the buildup but also reproduce themselves for each cell division, so that the buildup can go on with ever increasing multiplicity.

**Virus and Man**
So far we have described the structure and the life process of a bacterium. What about the other forms of life from virus to man, including all plants and animals? It is remarkable that some of the essential features are the same in all manifestations of life, in spite of the great variety of forms and species.

Let us look at the two extremes in this scale, from the virus (see figure 65) to man. A virus is much smaller than a bacterium. It consists of a nucleic acid contained in a coat made of simple proteins. It has none of the complicated proteins that produce amino acids and no ribosomes to put them together as proteins; it does not need them. A virus leads the life of a parasite. It can live and multiply only when it attacks some other cell. Clinging to the skin of the host cell, it sheds its nucleic acid into the cell, and there uses the amino acids and the ribosomes of the host (and the ATP energy sources of the host) for its own duplication and for the acquisition of a

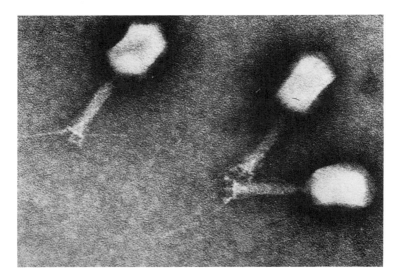

Figure 65
Three virus units as shown by an electron microscope. Magnification 180,000
times. (Courtesy of E. Kellenberger and J. Bron.)

new coat. In fact it works so fast that soon the cell is full of
newly made viruses, which burst the skin and kill the cell. (See
figure 66.) Thus viruses produce diseases. They attack certain
kinds of cells in our bodies and kill them by using up cell
material for their own reproduction. Naturally the nucleic
acid of the virus needs to "know" much less than the one of a
bacterium. It only needs to produce the proteins for its coat
and to reproduce itself. No other protein is needed, because
the host cell supplies everything else. It is no surprise that the
nucleic acid of a virus is very much shorter than the one
found in a bacterium. Its spiral ladder contains only about
100,000 steps. This short length fits very well into our picture
of the nucleic acid as the carrier of information. A virus is the
simplest living unit; it has the shortest DNA chain.

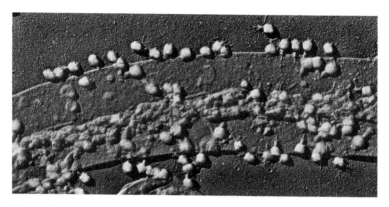

Figure 66
A bacterium cell sheds many virus units that were produced inside. Electron microscope magnification 48,000 times. (From E. Kellenberger and W. Arber, *Zeitschrift für Naturforschung* 106 (1955).)

Let us now go over the the most complicated living thing, man, and compare him with the bacterium. First of all, the bacterium is one single cell; man is an agglomeration of many cells of different kinds. But in spite of their variety, human cells are similar to the bacterium cell in many essential features. They are made of proteins—simple fiber-shaped ones for the skin and the structure, complicated globular ones for the performance of chemical reactions. They contain ribosomes to produce proteins and ATP molecules, which carry energy around. Last but not least, each cell contains DNA molecules. The DNA molecules are, as expected, very long—about a thousand times longer than the bacteria DNAs. The spiral ladders in human cells are several feet long when stretched out in a straight line. In actuality, however, the ladders are coiled and take up very little space in each cell. They contain several billions of steps! The enormity of this array, of course, has something to do with the fact that a human being is a more complicated organization than a bacterium.

What is most exciting, however, is the fact that the human DNA uses the same code triplets as the DNA in any other living entity, be it a virus, a bacterium, a plant, or an animal. The "dictionary" mentioned earlier is valid for all of them. The human cells, as well as the cells of any other living subject also, use the same energy source, namely the ATP molecules, be it for the synthesis of chemicals or for muscle contractions. Life in all forms makes use of the same trick.

Although all human cells contain the same DNA, they differ from each other in many respects. Each type of cell serves a different purpose. The skin cells protect the body, the stomach cells produce the chemicals needed for digestion, the muscle cells are able to contract and perform physical work, the cells of the retina in the eye are light-sensitive, and then there are the nerve cells.

The nervous system is perhaps the most important innovation in the progression from the bacterium to higher species. Nerves are long strands of special cells that, like telephone wires, transmit messages from one place to another, though at lower velocity and by a different mechanism. It is a somewhat different mechanism than a wire, but an electrical impulse, nevertheless. When light falls upon the retina, the stimulus is transmitted to the brain; when the skin is stimulated at some place, the message is transported to other places. Conversely, a stimulus originating in the brain goes via the nerves to any muscle of the body and produces a contraction. The brain itself is a complicated tangle of an enormous number of nerve cells, as many as ten billion, which are interconnected and arranged in a way that we do not understand yet. But this tremendous unit of nerve cells is able to react to the stimuli coming from the outside. It can think and feel.

There is an essential biological difference between the bacterium and man. When the bacterium grows and divides, it produces everything that is needed in one single cell. When a human being grows, very different cells must be fabricated— skin cells, muscle cells, bones, nerves, etc. Each part of the body must follow a different master plan.

This fact raises a most formidable problem that is not yet completely solved. According to our present knowledge, every cell contains a DNA molecule that possesses the whole information of the body; it could produce all the proteins ever used in man. Evidently none of the cells produce them all. How then does the cell know which part of the DNA master plan concerns its growth? How does it eliminate all other parts and make use only of the appropriate one?

Consider the problem of growth and development of the individual. We know that at the very beginning the embryo consists of one cell, which divides into a few more, each of them alike. But soon specialization occurs. Some of the cells develop into the spine, others into the limbs, others into the gastric canal, and others into the nervous system. Although they all have the same nucleic acid, they develop differently. How is this done?

We notice this peculiar selective ability of cells when we observe the healing of a wound. Somehow the cells adjacent to the wound "know" how to grow and to multiply in such manner and direction that the original structure is reestablished. Only those proteins are produced that fit the pattern and cause the desired structure to develop.

Very probably the explanation of this problem goes along the following lines. There are chemical substances in the cell called *repressors* that block those parts of the DNA that should not be used. It depends on the environment around the cell which parts of the DNA are blocked by repressors and which are left active. At the beginning of the development of the embryo, only those parts of the DNA are active that produce the early cells. Then the presence of the newly created proteins influences the repressors such that they leave other parts of the DNA active, thus creating a new set of proteins. We see that at each step of the development, different parts of the DNA become active. In each cell of the human body, only the part of the DNA corresponding to the specific needs of the particular cell is at work. The other parts are held inactive by the repressors. We do not know yet exactly how the environment of the cells influences the repressors to leave only those parts of the DNA unblocked that are relevant to what should grow in that environment, but such influences clearly exist. Experiments with various embryo cells have shown, for example, that a cell taken from a position where it was destined to become part of an animal's tail will develop as part of the animal's head if it is transplanted into the environment where the head is supposed to grow.

The order of nucleotides in the DNA is not maintained exactly, when a new individual is formed. The chain of molecules is so tremendous that small deviations here and there do not fundamentally alter the development of the individual; they cause only small modifications. Thus no individual is exactly like another. As a consequence of sexual reproduction, these modifications are mixed in every new generation. Children are basically like the parents, but different in details. Their DNA is a copy of about half of the DNA of the father and half of the mother. Here is a fundamental difference between living individuals and atoms. Two atoms of the same kind are identical in every respect—they are completely alike. Two living beings of the same species never are completely identical. The immensely long chain of the DNA provides many possibilities for variation within a given species, and it is the variations that make life so interesting and exciting.

## The Sources of Nutrition

The material that forms human beings and makes them grow—where does it come from? The virus solves this problem very simply. It is a parasite, and it forms its progeny inside the host cell, using up the amino acids and nucleotides produced by the host. The bacterium is more independent; it builds up the amino acids and nucleotides by means of special proteins, but it needs to be immersed in a nutrient solution of sugar and other chemicals. Man and animals must eat in order to live and grow. They eat living material, such as plants or meat. In this respect man and the other animals are less independent than a bacterium. We could not live off a sugar and ammonia supply because human and animal cells cannot synthesize some of the essential amino acids. We are parasites in this respect. We must get a good part of our supply of amino acids from other living material. The proteins in our food are broken up, in the process of digestion, into the amino acids of which they were made. The human and animal cells then use these amino acids in order to build up the proteins needed for their own growth and for chemical work.

As far as our energy needs are concerned, for the work of our muscles and the synthesis of proteins, we are doing the same as the bacterium. We, too, have proteins in our cells that can make use of the energy liberated by the "burning" of sugar, and we can store it in small portions in the quantum states of the ATP molecule. These molecules are the energy carriers of the body; they are taken up by a muscle when it contracts and does work, or by a cell using energy to produce new proteins.

We and all other living beings, including the bacteria, use up sugar by burning it to form $CO_2$ and water. We and the animals also need some of the amino acids to build up our cells. Where does it all come from? There must be a place where sugar is synthesized in order to make life possible and amino acids are made for animals. If there were no such place, living beings would soon have used up all the available supply.

Plants are the places where this synthesis occurs. The green color of plants comes from a chemical called chlorophyll, which is, next to DNA, the most crucial molecule for the existence of life on Earth. It is not as large a molecule as DNA, but it has an intricate structure with which it performs an important function. When exposed to the rays of the sun, chlorophyll makes use of the energy of sunlight and rebuilds energy-rich molecules, such as sugar, from their "ashes," carbon dioxide and water. Sun energy is transformed into chemical energy. Thus chlorophyll, with the help of sunlight, reverses the process of burning. The burning of sugar to $CO_2$ and water releases the energy needed in the life processes. The sugar must be replenished. In photosynthesis—that is the name for the action of chlorophyll under sunlight—the sun energy is used to transform carbon dioxide and water back into sugar. The chlorophyll molecules work in the cells of green plants. A plant gets the water from the soil, the carbon dioxide from the air, and the energy from the rays of the sun.

Sugar contains less oxygen than $CO_2$ and water. Hence, as a by-product, free oxygen develops when this process goes on. Most oxygen in the atmosphere was produced by plants when their chlorophyll did its sugar-producing job. We could not breathe were it not for the plants' constant work, producing oxygen.

Chlorophyll molecules are fabricated whenever a green plant grows. The plant cells are similar to the cells we already have described; they are, however, even more independent than the bacterium. Not only do they have proteins, which make all the necessary amino acids, but they also can fabricate chlorophyll, which then, with the help of sunlight, synthesizes sugar. In this way plants live and grow without a nutrient sugar solution and without "eating" living matter. All they need are light, carbon dioxide, water, and a few chemicals such as ammonia, which is found in the soil. In this respect they are ideal living systems.

The plants are the only kind of living matter that is "productive"; they make all their material from simple minerals with the help of light. All other forms of life are "destructive." They need the energy-rich material formed by plants, and use it to produce their own structures. Most bacteria only need sugar; animals and man are the worst offenders. They need not only energy-rich material like sugar but also organized material such as amino acids in order to build up their own cells. On the other hand we and the animals can boast of a much higher organization, vastly higher than in other forms of life. In particular, we must be proud of our nervous system, which gives us coordination between sense perception and motion, and last but not least, makes thinking possible.

What is life? From our point of view it is a manifestation of some special molecule structures that reproduce themselves continuously according to an established scheme. This scheme, though different in each species, is based on processes that are the same for all species. But we must keep in mind that these processes can occur only under very special conditions. The environment must be just right in order to allow atoms aggregate to form living matter. The temperature must be low enough so that heat motion does not destroy the complicated fabric of macromolecules. It must not be too cold either, since life is possible only if proteins and nucleic acids can perform their chemical synthesis; for this activity some heat motion is necessary. If the cell material freezes, all chemical work comes to a standstill. Sunlight must be available for the synthesis of energy-rich molecules, but not too much—the temperatures must remain moderate. Obviously, here on Earth, these conditions are fulfilled; the surface of our planet abounds with green plants and various strange combinations of atoms that we call living organisms.

Although many questions of human and animal development are unanswered, some ideas stand out clearly. Each species with all its organs, nerves, bones, and brain develops biologically from its germ cell according to the plan laid out in the nucleic acid macromolecule. Here atomic physics and life in its highest form are intimately connected. Each nucleotide in the long chain has its well-defined quantum state, which is the basis of its specific character. They are tied together by electrons in typical quantum patterns, which are stable enough to maintain the order of the chain in spite of the heat motions and other disturbing effects in the cell. Upon this order rest not only the development of the individual but also the propagation of the species. The stability of the quantum patterns in the DNA is the guarantee that the children are basically like their parents, that the species is maintained. The various forms of life are a reflection of the various ways of combining nucleotides in the nucleic acid. The constancy of these forms, the recurrence in each generation, is a reflection of atomic stability.

about the first moments of our world. Let us see what present day science can say about it.

We return to the discussion of the expansion of the universe that we began in chapter 2. As we reported there, the astronomers observed that the galaxies move away from us and from each other. The farther they are, the faster they move away. The totality of space expands, matter becomes more and more dilute, the distances between the galaxies increase steadily. These circumstances lead to the conclusion that way in the past the galaxies must have been closer together. Indeed there must have been a time when the concentration of matter was such that the galaxies were not separated at all. We can estimate the time elapsed since that moment from the present speed of expansion and find that this happened about ten to fifteen billion years ago. It is reasonable to assume that the expansion of the universe did not start when the galaxies were merged. Most probably this was already a late state of the development, preceded by earlier periods of much higher density of matter than we find in the galaxies today. Higher density is always accompanied with higher temperature and pressure; expansion to lower density has a cooling effect. Thus we assume that in those earlier periods the temperature and pressure must have been extremely high, so that matter was on higher rungs of the quantum ladder than we find it today in our own environment. Molecules were dissociated into atoms, atoms and nuclei were split up into their constituents; at the very early stages, even the protons and neutrons were in excited states or dissociated into their components, the quarks. That highly energetic, extremely hot and compressed initial state of the universe is often referred to as the "big bang." Space was filled with all forms of highly concentrated energy, such as light, neutrinos, particles and antiparticles, electrons, positrons, quarks, or whatever were the primordial parts of matter. The enormous pressure and energy concentration led to a forceful expansion, the birth of our present universe. One can estimate that it took only a few minutes for the universe to cool down from that primordial state until matter assumed forms that are better known to us, such as protons, electrons, and a few neutrons. After another short time, maybe a quarter of an hour or a little more, the temperature dropped sufficiently so that the neutrons formed helium nuclei ($\alpha$-particles) with some of the protons. At this point, the hot universe was filled with protons, electrons, and $\alpha$-particles, about one $\alpha$-particle to fifteen protons. It was still much too hot for the formation of atoms. It then took almost a million years to cool down so that the electrons and protons could get together to form hydrogen atoms without being torn apart immediately. At the same time, two electrons and an $\alpha$-particle also formed helium atoms. Then the universe consisted of a mixture of hydrogen and helium gas, but hydrogen was most abundant.

In the process of further expansion, the gas separated into large distinct patches, the forerunners of the galaxies. Those large masses of hydrogen gas with a few percent helium mixed in are the raw material of which stars are made, as we will learn.

In many ways, these ideas about the beginning of the world are rather hypothetical. But during the last decade, a very exciting observation was made that strongly supports the ideas of the big bang. It has been observed that the space around us is filled with a weak radiation of very long wavelength. What has this radiation to do with the big bang? It can be interpreted as some kind of echo, of an optical reverberation of the bang that still fills the universe today. The conclusion goes as follows. At the time of the big bang the universe was full of an extremely hot and intensive radia-

There is little fact to support these ideas, but there is not much to contradict them either, except that they cannot explain the existence of the three-degree radiation. It is true that the notion of matter being created out of nothing in space contradicts our ordinary ideas of conservation of matter and energy. However, the amount required to keep the universe from thinning out is so small that one never would observe it in a laboratory. All that is needed is one hydrogen atom per year created in each cubic mile of the universe; it would be impossible to observe these infinitesimals directly. These speculations about continuous creation of matter are of interest because they show that an expanding universe without beginning or end and with a constant matter density is at least logically possible.

## The Evolution of Stars

In our account of the evolution of the universe, we have reached the stage where it consists of large patches of hydrogen gas mixed with traces of helium. These patches of gas represent the early stages of galaxies before the formation of stars. How does the gas give birth to stars? We know a good deal about the development of stars from gas clouds. The process is also going on today, since the galaxies still contain clouds of hydrogen gas mixed with other traces. These clouds are either remnants of the original gas or are the products of star explosions.

We now get to the second part of the story of evolution of the world, how all things we see and live with have evolved from a cloud of hydrogen gas mixed with a little helium. A word of warning is in place, however. Even this part of the story of the universe is based upon rather unsafe conclusions and theories concerning the behavior of matter under abnormal conditions. It is not as speculative as the evolution from the big bang to the hydrogen-helium gas, but many of the statements are hypothetical, too, and are much less solidly founded than the contents of previous chapters in this book.

**The Hydrogen Gas**   We start with an enormous cloud of hydrogen gas. (The small admixture of helium is not important; we will forget about it in what follows.) The great Orion nebula is an example of such a cloud. (See figure 67.) It is visible because it is illuminated by neighboring stars. There is not much variety or much order in this cloud. Hydrogen atoms are moving at random, occasionally colliding with one another.

The cloud undergoes slow changes under the influence of gravity. It is true that the gravitational attraction between hydrogen atoms is extremely weak because of their small mass. If the cloud is very large, however, the combined gravitational effect of many atoms becomes important. Over long periods of time groupings do occur, and such a grouping represents a stronger center of attraction than a single atom. It therefore attracts more atoms and becomes an even stronger center, which in turn forces more and more atoms into its region. Finally one or a few large, dense clusters of hydrogen atoms are formed, and they grow bigger still until they have attracted most of the material of the cloud.

The gravitational force goes on pulling the atoms together, so that the cluster becomes smaller and denser. The atoms "fall" toward the center under the pull of gravity. During this fall they acquire speed; when they get into dense regions, they will collide with other atoms and transmit the energy of their motion to the rest of the material. So the contraction of the cluster causes the atoms to move faster and faster and collide with each other. Therefore the gas gets hotter and hotter. Gravitational energy is transformed into irregular heat motion. The cluster gets denser and warmer.

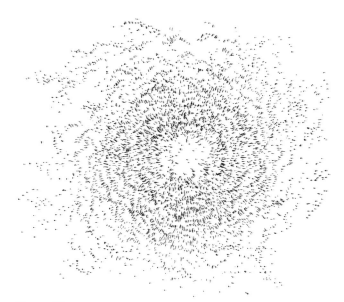

Figure 68
Hydrogen gas cluster.

sponding counterpressure generated at the center stops the gravitational collapse. The contraction ceases, and for a long time afterward the star remains the same size.

Hydrogen burning is a slow but efficient process. The heat produced by this fire works its way to the surface and maintains the brilliance of the star. It supplies the energy needed for constant emission of light from the stellar surface into space. It keeps the star going for a long time. Several billion years pass by before the hydrogen in the hot region at the center is depleted and turned into helium.

**The Third Stage of a Star**    When most of the hydrogen is exhausted in the central region, the hydrogen fire comes to an end and the counterpressure that had prevented gravita-

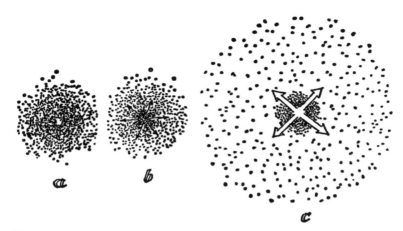

Figure 69
The first three stages of a star. (a) Hydrogen gas ball. (b) Hydrogen burning takes place at the center and develops counterpressure. Heat flows outward from the center. (c) Red giant with helium burning at the center, radiating enormous heat. It disperses the outside material over a large sphere.

tional contraction subsides. Then gravity does its work again, and the star contracts further. This is always connected with a rise in temperature, since the atoms "fall" inward and gain speed. There comes a moment then when the center reaches the temperature of several billion degrees, at which helium starts burning. This is the beginning of the third stage of the star. The helium nuclear fire comes from the formation of carbon in a fusion of three helium nuclei (as we saw in chapter 7). The helium fire burns rather fast and produces much more heat than the slow-burning hydrogen fire. The pressure of the heat at the center not only stops the contraction, but pushes the rest of the material of the star away from the center. The star then consists of a very hot and dense center, where the nuclear burning takes place, surrounded by a giant

highly probable that these heavy nuclei were created during
the process of explosion and then scattered over vast space.

A new "second generation" of stars is formed of original
hydrogen gases into which star explosions have mixed their
material. The development of these second-generation stars
is not very different from the first-generation stars, since the
admixture of nonhydrogen material is very weak. The gas out
of which they develop is still mainly hydrogen.

The sun is an example of a star that originated from a
hydrogen cloud contaminated with remnants of an exploding
star. It is now in its second stage of development, burning
hydrogen to helium. This slow and regular nuclear fire
supplies the energy with which the sun has radiated its warm
light steadily for several billion years. Had the sun originated
from a pure hydrogen cloud, it would contain nothing but
hydrogen and helium. The sun does consist mostly of hydrogen and helium, but when we investigate light emitted
from the sun's surface, we find traces of other elements, and
these elements testify to the fact that the sun is from a "later
generation" of stars; some of the material of which it is made
had been in another, earlier star.

**The Creation of the Earth**

When the sun was formed from the original contaminated
hydrogen cloud, some special process must have occurred
that placed little chunks of matter in orbits circling around
the sun. We know that the sun is surrounded by nine planets,
which are very much smaller than the sun itself (figure 70).
The genesis of these little chunks is of great importance to us,
since we live on one of them. We have only very vague ideas
of the mechanism that brought these planets into being. One
way to imagine their origin would be this: When the gas cloud
contracted and formed the star, small bits of cloud must have

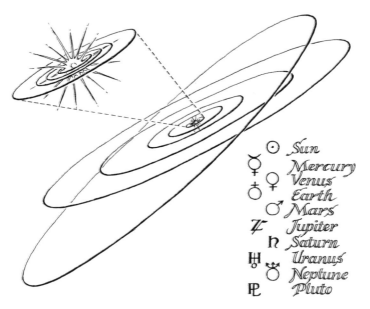

| | |
|---|---|
| ⊙ | Sun |
| ☿ | Mercury |
| ♀ | Venus |
| ⊕ | Earth |
| ♂ | Mars |
| ♃ | Jupiter |
| ♄ | Saturn |
| ♅ | Uranus |
| ♆ | Neptune |
| ♇ | Pluto |

Figure 70
Orbits of planets, and their symbols.

been left over outside the star. They gathered together by gravity and formed the subunits that now circle the sun as planets. Originally these left-behind parts consisted, of course, of the same material as the rest—that is, mostly hydrogen with a small admixture of heavier elements. Therefore some of the planets have the same constitution as the sun; this is true for the larger planets such as Jupiter, Saturn, Uranus, Neptune, and Pluto. It is not so for the smaller planets such as Mercury, Venus, Earth, and Mars. When they were formed, a separation of elements occurred; they lost a good part of their hydrogen and helium, and the remaining hydrogen was bound into molecules with other atoms, mainly

orderly arrangements. It starts with protons and neutrons forming nuclei in the center of stars, then atoms in space. Atoms aggregate to form molecules and, finally, liquids and solid are formed on a few planets where the temperatures are appropriate.

In this development only a very small part of the original hydrogen is transformed into other elements. Enormous amounts of hydrogen are needed to create the conditions under which a tiny part turns into more complex units. Stars must assemble, explode, and reassemble in order to turn a negligibly small part of the original stuff into the variegated substances and materials that we find on Earth.

How long did this development take? As we learned in chapter 2, the solar system originated about 4.5 billion years ago. The explosion of the stars that delivered the different atomic species must have occurred earlier; there is some indication that this happened about seven to ten billion years ago. The life span of an average star like the sun is estimated to be about five to ten billion years. It must have been about 10 to 15 billion years ago, therefore, when the first hydrogen cloud started forming the star that made our elements. This event must have taken place not very long after the beginning of the universe, the big bang—perhaps only a fraction of a billion years later. Much time and many materials were needed to create the substance of our world.

## The Development of Life
**The Beginnings**   We have followed the evolution of our world from the hydrogen cloud to the development of stars with planets. The planets are an agglomeration of matter in a more "advanced" form; they consist mostly of elements more complex than hydrogen. The earth, for example, is made up

of many elements. The bulk consists of heavy metals, mostly iron; the outer layers are rock and minerals; on the surface there is much water. The outside is covered by a layer of gas, the atmosphere.

The remainder of our story of evolution takes place beneath this protective layer of gas, on the surface of the earth. It is a story of further differentiation of matter, of the formation and propagation of complex units of matter that make up the living world on Earth.

Before telling the story, we repeat the words of warning that we expressed before. We know very little about the functioning of life, even under present conditions. Hence any statements concerning the origin and the early stages of life must be very tentative, since that origin occurred under very different conditions.

In facing this problem, as in facing the previous one of the development of stars and planets, we are in the position of explorers who are trying to draw a map of an unknown continent. The knowledge of the explorers is patchy; they have seen only small stretches of the coastline, followed a reach of water here, and found a river farther inland. When drawing the map they must rely on their imagination to fill in the unknown parts of the coastline, and they tentatively assume that the reach of water and the river are two parts of the same. It is the best they can do. A later map will show that they simplified and distorted the picture enormously and that, after all, their two bodies of water were not the same river. But generally, despite the errors, the explorers' outlines will be recognizable.

Our knowledge of the development of the world is patchy, too. We are forced to use our imagination at almost any step to fill in many unknown stretches. Many things I am saying

duced a variety of new chemical conpounds that were not present before.

Another way of producing new chemical compounds must have been the numerous electrical discharges that took place in form of lightening strokes when thunderstorms raged over the surface of the earth. Among these new compounds there certainly were some of the molecules that play a role in living structures, such as sugar, nucleotides, and amino acids. The formation of these molecules must have been a very slow process. The ultraviolet light and the electrical discharges first had to decompose molecules containing the necessary atoms, and then chance had to bring the atoms together in the right positions to form the new molecules. Simple structures must have been formed more frequently than complicated ones, because it is much more probable that a few atoms would get into the right position than that many of them would. Alcohol and sugar were formed by sunlight in much larger quantities than amino acids and nucleotides.

The formation was slow, but over many millions of years these substances did acccumulate. When formed at the surface of water, they sank to lower layers and were protected against breakup by ultraviolet sunlight. Today such accumulation would be impossible; amino acids or nucleotides would be quickly incorporated in living organisms or decomposed by oxidation with free oxygen in the atmosphere. The sterile conditions of the early periods permitted the slow accumulation. Thus it came about that the waters on Earth slowly began to contain small quantities of sugar and similar compounds and even smaller quantities of amino acids and nucleotides.

In the great oceans these molecules got lost easily and strayed far apart from each other. In smaller ponds or puddles, however, the concentration may have become not inconsiderable. Some nucleotides might even have joined together and formed a small chain of nucleic acid, and some amino acids might have joined to form a protein chain. The chains that would have been formed in chance encounters were not the chains that play a role in life; they were accidental combinations without any special significance or special chemical activity. One day one type of protein would have been formed and the next day another type.

The nucleic acids formed by the accidental joining of nucleotides had more lasting effects. We saw in chapter 8 that a nucleic acid chain of the DNA type can reproduce itself exactly by dividing into two halves, and that each of the halves then collects the correct nucleotides for the buildup of two identical full chains. Therefore if a nucleic acid chain is put into a medium containing nucleotides, the chain may produce more and more replicas of itself until all the nucleotides are used up. So if one chain should be formed by accidental encounter, this chain then would induce all other nucleotides in the neighborhood into forming chains of the same kind. Nucleic acids will produce replicas in an environment where nucleotides are available. In many respects this property is the basis of life, since it allows a complicated structure to reproduce itself under favorable conditions.

Nucleic acids can do more than just reproduce themselves. We learned in chapter 8 that they carry instructions for the formation of proteins. Always three subsequent nucleotides correspond to an amino acid. We have seen how these instructions are used by the ribosomes to produce proteins. Of course, at the beginning of life, ribosomes didn't exist; they were a later achievement in the course of evolution. Before that, probably, some simpler and less efficient mechanism

energy carriers that we met under the name of ATP molecules in chapter 8.

(d)  One serving as a coat or skin for the nucleic acid, a skin
that has small pores that let simple chemicals through, but
keep most chain molecules inside.

(e)  One that forms special molecules capable of synthesizing
sugar with the help of sunlight. (Chlorophyll is an example.)[30]

Let us now discuss the effects of these useful proteins. We
already have described the great usefulness of protein (a) for
nucleic acid formation. The protein (b) would strongly increase the amino acid supply, which previously was furnished
only by the slow method of production by ultraviolet light or
lightening discharge; hence it would enhance the formation
of any kind of protein by the DNA chains.

The protein (c) accelerates all chain-building processes,
since it provides suitable energy carriers that help to link one
molecule to the next in the chain. Before these ATP energy
carriers were available, the energy needed for the linking was
supplied by heat, a very unreliable and slow way to build a
chain.

The protein (d) serves a most significant purpose. Before
the formation of this protein, the body of water acted more or
less as one unit. The multiplication of nucleic acids made use
of the total supply of nucleotides. For example, in that fa-
vored body of water where nucleotide-producing proteins
were formed, any nucleic acid would be multiplied, not just
the one that carried the code for nucleotide production. A
specific nucleic acid, therefore, does not enjoy the advantages
of its talent to the exclusion of other nucleic acids. They all

30. We have simplified the situation by pretending that one protein only is
needed for each of these tasks. In fact each task needs a whole system of
several special proteins, but the trend of thought here still holds.

will multiply and use up the raw materials. But if the effective nucleic acid also can produce a skin, the productive protein and its products will be kept close, and the products will be available exclusively to this specific nucleic acid. Hence this one alone would be able to produce many replicas, and it would develop very much more rapidly than the others. Not only would the others be deprived of the increased supply of nucleotides, but the skin would keep the building materials close at hand when they were needed.

With the formation of a nucleic acid that can produce proteins of the types (a), (b), (c), and (d), life has started to exist. Here we have something closely resembling a bacterial cell. When such a unit is found in a puddle of water with sugar and other simple chemicals, it actually lives. Amino acids and nucleotides are produced within the unit, the amino acids are put together to form the necessary proteins by the nucleic acid and the nucleotides are used when the nucleic acid forms a replica of itself. When the unit becomes too big, it will burst, and each separate nucleic acid will form its own unit again. This bursting and reforming might have been the first and simplest way of cell division. It was a very wasteful one, since many substances were lost in the process. Nowadays a cell has a much better way to divide without loss of material.

Even this advanced chemical unit could not go on multiplying itself forever, since it "fed" on the simple chemicals such as sugar, phosphates, and ammonia. There was and is no shortage of phosphates and ammonia on Earth; they are simple low-energy compounds available in large quantities. But the supply of sugar was not unlimited. Sugar, a chemical compound of "high energy," was made by ultraviolet light or electrical discharges in small quantities only. When the sugar supply of a body of water was exhausted, our units no longer could multiply. The ones destroyed by external causes, such

the "mutation" of the structure of the nucleic acid. We use the term "mutation" for the following phenomenon: In the course of self-reproduction it is bound to happen that sometimes the replication of the nucleic acid chain is not exact. Changes will occur from time to time.

The changes we may expect will be of two kinds. First, errors occur in the process of replication. The new nucleic acid sometimes is not exactly the same as the old one. If the new form becomes unable to produce the necessary proteins, the unit in which the change has occurred will cease to develop. If, however, the new form of nucleic acid produces the necessary proteins in spite of the change, the alterations will be repeated in each replica, and, therefore, they will be maintained from then on in the progeny.

Second, nucleic acids may acquire additional groups of nucleotides and thus increase their length. After all, the chains of the first nucleic acids were quite short; they carried the pattern for a very few proteins only. Whenever a few nucleotides are added by some chance, these additions are reproduced from then on in the replicas. In most cases these additions are valueless for protein production. Over very long periods, however, it is bound to happen from time to time that they give rise to a better protein, or that the extension of the nucleic acid can produce an additional new protein that will help to use more efficiently the raw materials available for reproduction. Whenever this happens, this new type of unit soon replaces the old one, since it multiplies faster and therefore uses up all available material for its own replication.[31]

31. One specific mechanism for lengthening the nucleic acid chain deserves special mention. Imagine a nucleic acid chain that has been replicated but does not separate completely from the original molecule; the ends might still stick to each other. A chain of double length is formed. This new chain cannot produce more proteins than the old one did; it produces the same

It might also happen that the new proteins allow the units to multiply under different external conditions. For example, the old unit might multiply best in warm water, the new one in colder water; or the old one in the deep parts of the water, the new one in the shallow regions near the shores. Then the new types will not displace the old ones, but they will populate regions on Earth where the new conditions exist.

Here we have the process of natural selection. It is bound to set in whenever certain units have the capability of reproducing themselves and when the master plan of this reproduction undergoes arbitrary changes. These two factors, self-reproduction and mutation, work hand in hand. If the reproduction is unaffected by the changes, nothing much happens; the changes are bequeathed to the progeny. If multiplication is reduced by the changes, the units afflicted will die out. If the mutations give rise to units that reproduce more efficiently, these units will replace the old ones. Thus a slow development goes on toward units that are better adapted for multiplication under the existing conditions.

There is a characteristic trend in this development—the units are bound to become more and more complicated. They lose the simple features they had at the beginning of life's history. Most changes are steps toward higher differentiation, toward longer chains of nucleic acid, which produce more proteins with more specialized tasks. Hence from the moment when units exist that can form replicas of themselves, a development toward more and more complicated units is bound

---

ones, but twice over. However, it is less endangered when changes occur in further replications. Whenever a change does occur, the other half still remains intact, and it can produce the necessary proteins. Hence changes can be transmitted to the next generations that, without this doubling, might have died. Such accumulated changes may, at the end, lead to the production of new, additional, useful proteins, which may improve the processes of life.

Let us now look at the development that has occurred by natural selection. We start with a rudimentary plant cell; it contains the necessary proteins that make amino acids and nucleotides from sugar, phosphate, and ammonia. Let us assume that it also contains chlorophyll, which produces sugar with the help of sunlight. We then can regard it as a unicellular plant, a plant consisting of one cell. Most important, it contains the nucleic acid that carries the code for the fabrication of all its proteins. At the beginning, the self-reproduction of the rudimentary cell is a rather primitive and inefficient process in which the nucleic acid produces a copy of itself and the copy produces and assembles its own proteins and forms a new cell.

In the course of further development, more and more steps are added to the nucleic acid. It therefore carries the instructions for the formation of more proteins that help to make the cells more advanced and better organized than the first rudimentary cells. Indeed, the process of cell division has become so much better organized that none of the cell substances are lost in the division. It has become an elaborate mechanism, initiated and governed by the actions of suitable proteins.

Nor does growth stop here. Much more protection against disturbances from outside and much more favorable conditions for the multiplication of cells can be achieved if several cells combine and act together as one multicellular unit. The

---

(measure of disorder) must increase steadily. The increase of "order" in a living structure is always accompanied by a decrease of order in the sustaining physical environment. This balance is most important in the buildup of organic molecules in plants, which is done with the help of sunlight. For every molecule constructed, a certain amount of light energy is absorbed. Light is a highly ordered form of energy. Its absorption represents a large loss of order, compensating for the increase in order in the newly formed organic molecules.

unit can work more efficiently if the functions are divided among different cells. Some cells can serve as framework, some can collect the raw materials from the water or the ground, and others can grow where the sunlight is intense and serve mainly as sugar producers.

By adding one complication to the other, units such as our present plants were developed, consisting of millions of cells for many different purposes. In time the seas were filled with green algae and a green blanket of plant life started to cover the continents (figure 73). Once this cover of plants was established, new possibilities of life arose. Two essential things had changed. First, there was now available a plentiful supply of

Figure 73
A green blanket of plant life covered the earth.

sugar, nucleotides, and amino acids in the plants, a supply that renewed itself continuously by multiplication. Second, the sugar production by chlorophyll set free an enormous amount of oxygen gas. The atmosphere of the earth slowly filled with oxygen and the oxygen remained in the air because all losses from oxidation and other chemical reactions were constantly replenished.

Let us look at the effects of these two most important changes. Before the spread of plant life, it was most useful for a living unit to contain chlorophyll because the unit then could produce its own sugar. Sugar, since it was produced most inefficiently by ultraviolet radiation, was very scarce on Earth. After the spread of plants over the earth, however, sugar was plentifully available in the plants. The same is true to an even higher degree with respect to amino acids: These more complicated molecules were in very short supply before the spread of plants, but afterward the surface of the earth was covered with them.

Consequently at that stage of development, living units could exist that were unable to produce their own sugar or amino acid. They could develop easily and multiply by "feeding" on the supply of these substances in plants. This fact has most interesting consequences. Before the plant cover, any mutation that destroyed the nucleic acid's ability to produce amino acids and chlorophyll would make it impossible for the unit to multiply efficiently and the changed units would die out. But after the plant cover originated, such changes were not so dangerous; the unit could go on multiplying by thriving on the plant supply. Therefore many changes that earlier would have died out now were able to survive and multiply. This is why, after the plant cover, new kinds of living species developed; we call them animals. Freed from the necessity of producing fundamental chemicals, such as amino acids and chlorophyll, these new units developed their nucleic acid master plan in new directions. Multicellular units originated where the different cells had other functions too, such as locomotion and sensitivity to light and sound. They could move, see, and hear.

We must keep in mind how slow this development has been. It stretches over one or two billion years. The changes come about by an accumulation of mutation effects. It takes a long time before an accidental change or increase of the nucleic acid chain leads to a useful addition to the master plan. Here nature has introduced an ingenious new and efficient way of accelerating the process of development: Two units unite before replication so that a mixture of their nucleic acids is replicated. This new way, sexual replication, has the great advantage of combining new, successful trends that occur in each individual. It accelerates the development of better-adapted units.

The mixing of two slightly different master plans at every replication produces many new combinations. Indeed, exactly speaking, no two individuals will have exactly the same master plan inscribed in their nucleic acids, except in the case of identical twins. Thus sexual mixing produces a large pool of somewhat differing master plans within a population. Then the population is ready to adapt quickly to a changing environment or create more efficient units. There are always some variations of the master plan available that are adaptable to some new situation or capable of some improvement in performance. They will win out and multiply faster than others. This is the reason why sexual reproduction is the most common form of replication among more complex units, whether they are plants or animals.

It is a most important fact in the evolution of living structures that changes inflicted or acquired during their lifetime are not inherited. A change in body structure inflicted upon an individual will never be inherited by its offspring. We can cut off the tails of all members of a group of animals and keep cutting off the tails of the offspring, but the newly born will always have tails. The reason is obvious. A change inflicted on the body structure has no effect on the nucleic acids in the cells, which contain the blueprint of the new individuals. As long as the tail is planned in the blueprint, it will develop in the offspring, regardless of what has happened to the parents' tails. This fact contributes significantly to the continuity of evolution. Temporary alterations do not affect the blueprint. Only slowly, by small variations in the structure of nucleic acids, and by mixing the nucleic acids of two individuals in the sexual process, does nature change the constitution of living structures.

Let us return to the second change that plant life introduced—oxygen in the atmosphere. Recall that the building up of proteins and nucleic acids needs energy. The energy was provided by certain proteins that regulate the burning of sugar to carbon dioxide and water and can store the energy in small packages within the ATP molecules. The burning of sugar without ample supply of oxygen is not easy. There are oxygen atoms contained in the sugar molecule itself that can be used for the burning. This type of burning, which uses the oxygen in the sugar, is called fermentation; it is an inefficient way of getting energy from sugar. When free oxygen became available in the atmosphere, it was much easier to burn sugar in the cell and store energy in the ATP molecules. New units originated that made use of atmospheric oxygen in their energy production. This led not only to a much faster growth of new cells, but also created an energy surplus within the units that could be used for moving parts of the unit. Muscles were developed, and they caused the extremities to move and perform work for locomotion and for gathering up food.

In those large multicellular units we call animals, the oxygen of the air could not penetrate easily into the body cells. Therefore the following change in the master plan led to much better-adapted units. There developed a system of arteries in which a liquid containing special red cells is pumped throughout the body. These red cells absorb free oxygen easily and transport it to all the body cells, which need it for energy production. The absorption of oxygen occurs in certain tissues—the lungs—which are constantly filled with fresh air. Thus the animals with blood circulation could make much better use of oxygen for their energy supply.

But the greatest step forward in this trend for better coping with environment was the development of the nervous system. This is a special combination of interlocking cells capable of transmitting stimuli from one part of the unit to the other. Thus, coordination became possible between the functions of different parts. The most important innovations made possible by the development of the nervous system were the sense organs. They are special cell accumulations that are sensitive to messages from the external environment such as light, sound, pressure, smell, etc. The messages received are transmitted through connecting nerve cells to other parts of the unit so that the unit is able to coordinate locomotion and other reactions to the outside conditions. As a result, the units could react on changes in the environment in many ways that were most useful for the protection of the individual and for the acquisition of food. The structure could move toward light; it could recognize food by its smell or its shape; it could avoid danger by moving away or by protecting itself when large objects approached. Our unit acquired what we call a "behavior."

The development of a nervous system was so useful and effective that any mutation or sexual combination leading to a larger or more intricate nervous system gave rise to increasingly successful units. Thus a continuous evolution toward an increase in nerve cells began, and led to the formation of a brain. This organ is an accumulation of a large number of interconnected nerve cells capable of storing the effects of the stimuli that the unit has received. The storage was the beginning of what we call memory. An action that previously has had good results with respect to food intake or avoidance of pain is kept in memory and repeated readily if similar circumstances recur. Obviously the ability to "remember" such situations was an enormous asset for our units and helped their struggle for survival under difficult conditions. It supplied the ability to learn from experience.

At the beginning, such memory and learning mechanisms were not very complicated. With modern electronic equipment one can easily construct a device with a "nervous system" that remembers past situations and determines its actions on that basis. A machine controlled by a modern computer may serve as an example. A system of interlocking nerve cells is in many ways equivalent to a system of interconnected electronic vacuum tubes or transistors. A device with a few thousand transistors can perform most impressive acts of remembering situations and avoiding them later on. But, in fact, the brain of even an insect is a more complicated device. It contains ten to a hundred thousand nerve cells. The human brain has as many as ten billion; it is infinitely more complex than any man-made computer.

The event of brain formation is an important step in the development of life. Before this event a living unit and its reactions to the outside world were completely determined by chemical structure. After the event, the reactions of the unit depended not only on its structure but also on its previous experience. Its behavior was determined not by the nucleic acid master plan alone but also by what the unit experienced in the course of its life. The individual unit was formed not only by its biological development from the nucleic acid but also by the effects of the environment on its behavior.

In the course of the development of the brain, the role of memory and acquired experience slowly became more important. Nerves transmit stimuli from one part of the body to the other; when they are suitably interlaced, they also can store information and transform it into concepts that later on may cause new actions. The enormous advantage of this mechanism put a high premium on the development of complicated nerve cell assemblies. Nucleic acids that produced enzymes capable of stimulating the growth of those assemblies led to the development of successful living units. Thus animals with some kinds of brains spread over the earth.

Let us keep in mind, however, that behavior based on learning and memory is only a very small part of the behavior pattern. Most of the behavior of primitive animals was predetermined. It was developed according to the nucleic acid master plan. It was inherited, as we say. Birds build their nests, feed their young, and migrate south in the winter by instinct. These behavior patterns are not learned; they are inborn. The nerve complexes causing these actions are already preformed in the growing body. The reactions acquired by learning are few. Birds learn certain ways of twittering; some higher animals learn certain hunting tricks. Most important reactions of animals are inborn, however, as is shown by the fact that in most species newly born animals are able to live normally when raised without contact with their own kind. Hence both the body structure and the social behavior of each individual are governed by the code within the cell. Shape and a large part of behavior are predetermined in the nucleic acid. They are repeated in each new generation and change only if different combinations of nucleic acids are formed by mutation or by sexual mixing. Both the body structure and most of the behavior patterns change equally slowly. Ants and bees have the same social structure as long as they exist as the same species, and this existence extends over many thousands of generations. The behavior pattern of higher animals is somewhat more flexible and adjusts more easily to different exterior conditions, but, on the whole, it does not change much faster than the body structure, which is determined by the nucleic acids.

**The Evolution of Man**    In our tale of evolution we have reached the point where something new is beginning to develop, brought about by a simple increase in quantity and diversity of the cells making up the nervous system. It happens often in the material world that an increase in quantity at a certain point gives rise to deep qualitative changes.

Let us look at an example of quality from quantity. Let us consider an open vessel filled with water in a room. When the temperature is below the boiling point, an equilibrium is established in which a certain number of water molecules per second evaporate from the surface, and the same number per second return from the water vapor in the air and condense at the surface. The water in the vessel remains seemingly undisturbed, in equilibrium with the (moist) air. When we raise the temperature but still keep it below the boiling point, there is only a quantitative change. There are just more molecules per second leaving and returning to the surface. If the number of molecules evaporating and condensing per second goes on increasing, however, a point will be reached where the returning molecules can no longer keep up with the leaving ones. When the temperature reaches the point at which the evaporation cannot be compensated for by condensation, the water is transformed completely into vapor. In other words, it boils away. To the onlooker it may seem that at the boiling temperature something special happens to the water. This, in fact, is not so; evaporation also occurs at lower temperatures. The decisive change is in the relation of the water to the surrounding air. At the boiling point, the air can no longer replenish the molecules lost by evaporation; so the evaporation, which was "harmless" to the water below the boiling point, "destroys" the water above this point.

We can observe a similar phenomenon in a solution of salt in water. If the concentration of salt is below the saturation point, the solution looks clear and no deposit is formed. Actually, however, very tiny agglomerations of salt molecules may form on the walls of the vessel, but the deposit is dissolved immediately. If the concentration of the solution is increased beyond the saturation point (for example, by boiling off some water), the speed of formation of deposit surpasses the speed of redissolution; salt crystals begin to form in most beautiful patterns. Again, it would seem to the onlooker that at this point the solution has acquired a creative ability to give birth to special crystal structures. Actually this is a quantitative relationship; the balance between deposit and dissolution has changed, more is deposited than dissolved.

Let us return now to the evolution of the nervous system in animals. We know that the nervous system enables animals to adapt themselves to their environment with the help of their sense organs and their memory. In fact we know that animals "learn" from experience, and this learning capacity is an important factor in survival. Yet a large part of animal behavior is based upon "instinct"; it is part of the biological inheritance.

When man evolved from the animal kingdom, something new must have happened. We contend that this new element is based solely upon a quantitative difference in the nervous system. By an increase in this system, nature established a new type of evolution that broke, and will break, all rules established in the previous evolutionary periods.

The elements of the new evolution are all present in the animal world: memory and learning and perhaps even the formation of concepts and ideas. Higher animals also are able to use their brains to draw conclusions from their previous experiences to reason out consequences of actions without having to perform them. To some extent, they can think of what would happen under certain conditions and prepare their actions accordingly. Have you ever observed a dog or a cat finding its way to the kitchen by some detour when the doors of the direct access are closed? Only, as in the salt solution below the saturation point, concepts, ideas, and conclusions of animals are yet too weak and too few to accumulate and develop. The attempts at learning and thinking in the animal world are mostly "dissolved" with the death of the individual. When man evolved, the constant increase in the complexity of the brain and the nervous system reached a point at which death of an individual no longer eradicated the gains that memory of experience had established.

The development of language and memory enabled an adult individual to tell a younger person about his experience and his reasoning, and the pupil could act as if he had had the experience himself, or had carried through the reasoning. The workings of the brain became complex enough to provide for vicarious experience and reasoning, to enable man to pool the experiences and the thinking of several individuals, and eventually to accumulate experiences and thoughts from generation to generation. This accumulation was made possible by the development of concepts, mental constructions, abstract ideas, and many other methods of formulation and transmission of thought, such as speaking, writing, and painting. The period of parental care of children is much longer in humankind than most animals; this provides much more opportunities for passing on experiences and thoughts. The difference between man and animal is analogous to the boiling and saturation phenomena. When experiences collected by the species as a whole become more numerous than experiences lost through the deaths of individuals, a new process begins: the formation of "tradition."

At this point evolution has overcome the barrier against inheritance of acquired properties. As long as parents cannot transmit their experience to their offspring, the behavior of each new generation is based exclusively upon biological inheritance; it is based upon what is inscribed in the master plan contained in the cells. The situation is not changed even if there is some transmission of experience from one generation to the next. As long as the sum of experience lost by death is greater than, or as great as, the sum transmitted to the next generation, there is no accumulation of experience. The behavior of each generation is essentially the same and is dictated by biologically inherited properties. But if the transmission of experience between generations is large enough to cause an accumulation, the young will learn from the failures and successes of the elders, and newly acquired behavior patterns will be "inherited," not via the nucleic acids but by word of mouth or by written records.

At this point a completely new form of evolution has begun. The behavior pattern changes much more rapidly than the biological changes in the body structure or the changes in the natural environment, both of which determine the behavior pattern in the animal world. The changes in body structure are bound to changes in the nucleic acid chains, which occur rarely and slowly; the changes in the natural environment also are rather slow. In the new evolution, the changes in the

behavior pattern are much faster; they become established
when a new way of behavior is found and transmitted by
tradition to the following generations. For example, man has
changed from a hunting animal to an agricultural one, from a
cave dweller to a city builder. He has developed his toolmak-
ing capacity from the carving of pointed stones to the building of machine factories. All this development took place in
time intervals infinitely shorter than the periods in which
biological or environmental changes have occurred—for
example, the time interval during which man evolved from
apelike animals. The large brain capable of thinking, the
formation of concepts, the use of language and later writing,
these bring about an accumulation of experiences that are no
longer lost when an individual dies but are developed further
with every new generation.

Once the critical number of nerve cells is reached and this
stage of development is attained, the further course is set and
will develop at a constantly accelerating pace. Again the
analog of crystal formation in a saturated solution of salt is
relevant. Crystal formation starts best from surfaces of other
crystals. The first one formed has no such surface available, so
it must take a relatively long time to form. But the next structures are formed at the surfaces of previously formed crystals.
This availability makes for a rapid increase in the speed of
formation. The greater the number of crystals formed, the
greater are the opportunities for new formation. The same
principle applies, then, to the formation of tradition. At the
beginning, when mankind first acquired the possibility of de-
veloping tradition, the formation was very slow. Once started,
however, it grew with increasing vigor and differentiation.

Tradition takes forms that are not always favorable to
the species. If, however, measures are found that are
favorable—for example, agriculture, the exploitation of met-

als, and so on—these measures initiate a new way of life within a few generations and bring about that sudden change in behavior that is typically human.

Science is just one of these new measures or attitudes that grew from the accumulation of ideas and experiences. It took many generations to disentangle the vast number of observations, to separate apparent connections from real ones, to distinguish superstition from scientific fact. But once a systematic method for recognizing facts was found, the scientific revolution of the last 500 years could get under way. There is no doubt that science constitutes an important step in the new kind of evolution that began with the formation of tradition.

The new evolution by tradition is not only much faster than the previous evolution by mutation and sexual mixing of nucleic acids, it is based upon a very different mechanism, namely the transmission of ideas, concepts, and value judgments instead of nucleic acid chains. There are many traits in the behavior of animals and men that are superficially similar, such as social structure, altruism, domination of one type of individuals by others. But in the animal world, we are dealing with a behavior that is dominated by biological heredity and only slightly altered by learning and adaptation. In the world of humankind, the behavior is dominated by the accumulation of tradition and experience, which is based primarily upon learning and adaptation. True enough, this accumulation is limited and often influenced by biological necessities and hereditary traits transmitted by the nucleic acids. But the development of human behavior is of fundamentally different character compared to the development of animal behavior. Some analogies between the behavior patterns of the animal world and human behavior may be quite misleading. One must be very careful in transferring the results of animal sociobiology to human societies.

Again our analogy with the salt solution is instructive. When the solution is not saturated, the salt forms only very small crystals on the surface of the vessel that quickly disappear. The "behavior" of the salt is dominated by its properties, such as dissolving in water and crystallizing only occasionally and in very small quantities. When the solution is saturated, however, many large crystals are formed and they, in turn, cause the formation of more crystals. The "behavior" of the large crystals is dominated by the structure of the existing crystal surfaces and by the shapes and patterns of the previously formed crystals.

So far, of course, it is only the pattern of behavior and thinking that is transmitted from one generation to the other. The body structure is still reproduced in the old-fashioned animal way of propagation, and this leaves it unchanged for many generations. But who can tell? Nobody can exclude positively the possibility of a development like that portended in Aldous Huxley's *Brave New World*. It may become possible to change at will the nucleic acids that determine the development of the species. Our knowledge of the mechanism of propagation is still very limited, but it grows dangerously fast, and human interference with the hereditary structure of germ cells is not altogether out of sight. If this aim is achieved, the planned inheritance of desirable properties of the body will be within reach.

Even without having attained this ambitious aim, the new evolution by tradition has left its mark on the planet and interferes everywhere, in an ever-increasing way, with the mechanism of the previous type of evolution. Man creates new races of animals by crossbreeding and purposeful selection. Man changes his environment by planting crops and building cities and industries, and the changes are much faster than any natural changes. The times are over in which nature alone developed its own forms, slowly, by trial and error, undisturbed over many generations. No longer do we rely on geologic events to change our environment, or on chance to produce mutations and new nucleic acid combinations, with man as a happy onlooker. We now take it upon ourselves to develop nature and our own species. This is an arduous task, full of pitfalls and responsibilities. We assumed this burden only a short time ago, and nobody should be astonished if we blunder now and then. After all, nature blundered in the previous evolution, when mammoths and dinosaurs acquired larger and longer dimensions until they were given up as dismal failures. We must proceed by trial and error, just as nature did. The pace of the new evolution by tradition, however, is infinitely faster than that of the old evolution by inheritance. Mistakes are punished immediately and cause tremendous suffering to the perpetrators and their offspring. We are responsible ourselves for what happens, and we cannot blame nature for it. We are no longer in the Garden of Eden.

But isn't man himself part of nature? The tradition that mankind has accumulated, the ideas, concepts, myths, and religions, are all effects of nature's influence on man in many ways. They originated through man's reaction to natural events, the behavior of his fellow men, the hardships of life in a difficult environment. Our bodies and the bodies of animals were shaped in the long and slow process of natural selection; they bear witness to the conditions in which life developed during billions of years, when only those nucleic acids were able to survive that gave rise to a well-adapted unit. The tradition of human thinking and behavior is also a product of the effects of the environment on man, this time on the brain and not the nucleic acid. It evolved within the comparatively short time of a few million years.

# Epilogue

Our story of evolution has reached the present era. We have seen how life and man evolved from the original hydrogen gas, or better, how we believe today it may have happened. We gave a highly simplified account of it in order to emphasize the essential trends.

It is a development from the simple to the complicated, from unordered chaos to highly differentiated units, from the unorganized to the organized. This trend, however, is not shared by all matter in the universe; the more developed areas are much smaller than the less developed ones. Only a small part of the hydrogen cloud is able to form stars; in a small part of the star (the inner part), hydrogen is transformed into heavier elements. Only a small part of these elements is ejected into space, and only a small fraction of the ejected matter assembles in the form of a planet near a star. A small fraction of all planets is sufficiently near but not too near to the star, so that water stays liquid and chemical reactions can occur. Only a very small part of matter on these planets forms the long chain molecules that are the basis of life, and only a small part of all living matter develops a brain.

Every step to higher differentiation in this development requires an abundant amount of less differentiated material. The nuclear oven in the center of the star would not be hot enough for the production of elements were it not surrounded by vast amounts of hydrogen. Life on Earth requires radiant heat that can be supplied only by a much larger body, the sun—whose material is in a more primitive stage—since no molecule can exist at high temperature.

It is often said that science has displaced man and his earth from the center of the universe, where man fondly believed himself to be, and has relegated him to some unimportant place. Our sun is only a small and undistinguished starlet in a corner of the enormous expanses of our galaxy, with many other stars like it. What is more, there are probably quite a number of other stars with planets where life has developed.

These might be depressing thoughts for some, but they may also have a different significance. The vastness of the universe, the billions of stars and the space between them are necessary conditions for the development of matter from simple, unordered particles, to atoms and molecules, and finally to the large aggregates that form animals and sentient beings. The spots at which matter acquires more differentiated shape are very few and selected. They must be considered as the most developed and most outstanding parts of the universe, the parts where matter was able to make fuller use of its potentialities. We find ourselves, therefore, in a very privileged and central position, since our Earth is one of these spots. There might be other places where the development has gone much further even than here, but on the earth's surface life has developed and produced a thinking species. Nature is reflected in the thoughts of these beings.

It is not a simple reflection. In man's brain the impressions from outside are not merely registered; they produce concepts and ideas. They are the imprint of the external world on the human brain. Therefore, it is not surprising that, after a long period of searching and erring, some of the concepts and ideas in human thinking should have come gradually closer to the fundamental laws of this world, that some of our thinking should reveal the true structure of atoms and the true movements of the stars. Nature, in the form of man, begins to recognize itself.

Joyfully a patient lover
Mankind's eager spirit strove
To unravel and discover
How creative Nature wove.
It revealed that one eternal
Essence permeated all,
In the grains of shell and kernel
At the heart of great and small
Always changing and yet holding
Things together far and near
Shaping, endlessly unfolding
To behold it we are here.

Johann Wolfgang von Goethe
(Translated by Douglas Worth)

# Index